An Annotated Catalog of the Type Material of *Aphytis* (Hymenoptera: Aphelinidae) in the Entomology Research Museum, University of California at Riverside

An Annotated Catalog of the Type Material of *Aphytis* (Hymenoptera: Aphelinidae) in the Entomology Research Museum, University of California at Riverside

Serguei V. Triapitsyn and
Jung-Wook Kim

UNIVERSITY OF CALIFORNIA PRESS
Berkeley • Los Angeles • London

University of California Press, one of the most distinguished university presses in the United States, enriches lives around the world by advancing scholarship in the humanities, social sciences, and natural sciences. Its activities are supported by the UC Press Foundation and by philanthropic contributions from individuals and institutions. For more information, visit www.ucpress.edu.

University of California Publications in Entomology, Volume 129

University of California Press
Berkeley and Los Angeles, California

University of California Press, Ltd.
London, England

Library of Congress Cataloging-in-Publication Data

Triapitsyn, Serguei V., 1963–.
  An annotated catalog of the type material of *Aphytis* (Hymenoptera: Aphelinidae) in the Entomology Research Museum, University of California at Riverside / Serguei V. Triapitsyn and Jung-Wook Kim.
    p.   cm. — (University of California publications in entomology ; v. 129)
  Includes bibliographical references.
  ISBN 978-0-520-09867-1 (cloth : alk. paper)
  1. University of California, Riverside. Entomology Research Museum—Catalogs.
  2. Aphytis—Type specimens.{ems}3. Aphytis—Catalogs and collections. I. Kim, Jung-Wook, 1968–. II. Title. III. Series.
QL568.A55T75 2008
595.79074'79497—dc22

                                                                    2008025296

The paper used in this publication meets the minimum requirements of ANSI/NISO Z39.48-1992 (1997) (*Permanence of Paper*).

CONTENTS

Acknowledgments

This project was made possible by a grant from the United States National Science Foundation (NSF) DEB 97-28626 to John M. Heraty, John D. Pinto and Serguei V. Triapitsyn. Jung-Wook Kim's participation on this project was partially sponsored by an NSF PEET grant DEB 99-78150 to John M. Heraty and John D. Pinto.

We thank Michael S. Schauff (United States Department of Agriculture) for the loan of primary types of *Aphytis* from the National Museum of Natural History, Washington, D.C. The late Mike Rose (Bozeman, Montana) facilitated their transfer and also made arrangements for a donation of many hundreds of *Aphytis* specimens, including some paratypes, to the Entomology Research Museum, University of California at Riverside. Gerhardt Prinsloo (South African National Collection of Insects, Plant Protection Research Institute, Pretoria, South Africa) loaned the type material of some South African species of *Aphytis* to Jung-Wook Kim. Terry Nuhn (National Museum of Natural History, Washington, D.C.) made many useful comments on the manuscript. John S. Noyes and Andrew Polaszek checked for presence of the types of several species of *Aphytis* in the Natural History Museum, London, England, UK.

John M. Heraty and John D. Pinto provided valuable input in the project at the time of grant preparation and also oversaw its progress; John Heraty and Douglas Yanega reviewed an earlier draft of the manuscript. Technicians involved in various stages of this project were Vladimir V. Berezovskiy (remounting of almost all non-type and some type material of *Aphytis*), Marina Planoutene (remounting of most of the type specimens), as well as Jeremiah George and Kaylan Le (databasing and labeling of specimens), and Jason Le (initial processing of specimens).

Note: Present addresses of authors: Dr. Serguei V. Triapitsyn, Entomology Research Museum, Department of Entomology, University of California, Riverside, CA, 92521, USA (serguei.triapitsyn@ucr.edu); Dr. Jung-Wook Kim, c/o Department of Entomology, North Carolina State University, Gardner Hall 2323, Box 7613, NCSU Campus, Raleigh, NC, 27695, USA (argids01@gmail.com).

Abstract

An annotated catalog of type material of the genus *Aphytis* Howard (Hymenoptera: Aphelinidae) in the collection of the Entomology Research Museum, University of California, Riverside, California, USA (UCRC) provides information on 75 valid species. Curation of the UCRC *Aphytis* slide collection involved remounting 7,390 specimens from Hoyer's medium into Canada balsam (a single specimen per each slide). This included 7,292 specimens that were remounted, labeled, and databased and 98 specimens that were remounted and labeled but not databased (the types of the South African species received on loan from the South African National Collection of Insects, Plant Protection Research Institute, Pretoria, South Africa). Among these are 309 primary type specimens (i.e., holotypes, syntypes, lectotypes, and neotypes) belonging to 51 nominal species, 2,473 secondary type specimens (i.e., paratypes, allotypes, and paralectotypes) of various nominal species, and 4,608 non-type specimens of many species of *Aphytis* from more than 50 different countries. Numerous nomenclatural and other mistakes (such as improper or coded labeling) in the treatment of the type specimens of the *Aphytis* species are corrected; several primary types are returned to the appropriate depositories, and many remounted secondary types are donated to other major taxonomic collections of Aphelinidae. Lectotypes are designated for *A. africanus* Quednau, *A. coheni* DeBach, *A. cylindratus* Compere, *A. equatorialis* Rosen and DeBach, *A. fisheri* DeBach, *A. holoxanthus* DeBach, *A. immaculatus* Compere, *A. lepidosaphes* Compere, *A. lingnanensis* Compere, *A. melinus* DeBach and *A. taylori* Quednau.

viii

INTRODUCTION

This catalog has been prepared in part to report the results of the NSF-sponsored project that dealt with salvaging the world's largest collection of *Aphytis* Howard (Hymenoptera: Chalcidoidea: Aphelinidae). The majority of specimens, including the types, had been placed in a temporary, water-soluble mountant (Hoyer's) and were in danger of degradation. These specimens were mounted on 4,380 slides and were either dry or in various stages of deterioration (Figs 1-8). Prior to this project, the Entomology Research Museum, University of California, Riverside collection (hereafter UCRC) contained more than 30,000 slide-mounted specimens of *Aphytis* identified to 87 valid species (Table 1), including 41 primary types (excluding primary type material of 56 species on permanent loan to the National Museum of Natural History in Washington, D.C. [USNM]) and about 1,500 paratype specimens. This collection largely consisted of material obtained (mostly reared) by some of the foremost researchers in biological control (H. Compere, P. DeBach, S. Flanders, P. Timberlake, and others), and represented almost 100 years of collecting from at least 73 countries around the world; many of the specimens are vouchers from cultures of the species imported into the University of California, Riverside, quarantine facility and released in the United States since the 1930's.

Aphelinids are cosmopolitan in distribution and occur in all terrestrial habitats. With a body length ranging from 0.5 to 2.5 mm, these parasitoids are solitary or gregarious, endophagous or ectophagous koinobionts. *Aphytis* spp. develop exclusively as primary ectoparasitoids of armored scale insects (Hemiptera: Diaspididae). Information about morphology, biology, ecology, systematics and utilization of *Aphytis* spp. was presented in Rosen and DeBach (1979). Many species of *Aphytis* have figured prominently in biological control programs directed against injurious armored scale pests on citrus, olive, figs, coconut and other economically important tree crops (Rosen and DeBach 1979, Rosen 1994).

Unfortunately, the initial physical condition of most of the specimens in UCRC could be described as very poor, primarily due to the use of an improper mounting technique, described in Rosen and DeBach (1979). Their choice of Hoyer's as a mounting medium and of Zut® as the primary ringing compound was disastrous. Most of the Hoyer's mounts, especially those ringed with Zut®, were completely or partially dry (Figs 1-5). A few slides ringed with Glyptal® (Fig. 6) were usually in slightly better condition but these also have the potential for future breakdown (Upton 1993). In addition, Hoyer's medium often becomes dark with age (Figs 7, 8). Upton (1993) demonstrated that water-soluble media such as Hoyer's are not suitable for mounting specimens intended for long-term storage and use in taxonomic study. Earlier, Schauff (1985) criticized the mounting technique used for the *Aphytis* spp. type material and pointed out that their proper storage and curation would be difficult.

The number of slides that required remediation is summarized in Table 1.

Table 1. Total number of slides/specimens of *Aphytis* species at UCRC in 1998 that needed to be remounted from Hoyer's into Canada balsam at the start of the project.

Aphytis spp.	Primary Total types	Other type material	Non-type material
Determined	15/41 3780/25224	332/1489	3433/23694
Undetermined	n/a 600/3000*	n/a	600/3000*
Total	15/41 4380/28224	332/1489	4033/26694

*Estimating an average of 5 specimens per slide (this ratio was about 6.67 per slide among the determined species).

In addition, the primary types of the species described by P. DeBach and D. Rosen, which are on permanent loan from UCRC to USNM (Schauff 1985), also required remediation. These were received from USNM via M. Rose. Thus, almost all of the type specimens both at UCRC and USNM (except for the few types already mounted in Canada balsam, such as most of the species described by H. Compere) needed to be completely remediated and curated. Two species, *A. lingnanensis* Compere and *A. melinus* DeBach, together comprised about 7,000 specimens; of these, only a part was selected for remediation because there were many duplicates in the collection. However, most of the valuable voucher specimens as well as rare specimens from exotic locations were selected for remounting. This task was even more difficult because of other major mistakes made in the course of original mounting, as follows: 1) poor label data (Fig. 9) or occasional presence of only code (catalog) numbers or Quarantine Shipper and Receiver (S&R) numbers; 2) presence of several (often up to 100) specimens on the same slide under one or several coverslips (Fig. 9); 3) presence of several species under the same coverslip (Fig. 10); and 4) some labels that peeled off and became disassociated from the slides.

Other associated problems with *Aphytis* specimens are as follows.

Loans. No strict record-keeping policy of loan activity was maintained for the collection administered by the former Division of Biological Control at the University of California, Riverside (UCR) (1962-1989). Some of the specimens were received on loan from other institutions (no accompanying loan documents could be found). In a number of instances UCRC specimens were sent to various individuals on loan but have not been returned, and only a few loan records can be found. As part of the curation activities associated with this project, specimens belonging to other institutions were identified and most of them properly returned (we did not receive any response from some of the collections). A bar-coding system was implemented to help track specimens of *Aphytis* in future. Prior to remounting, many of the slides contained as many as 60-100 specimens. Bar-coding ties specimens (often of different sexes and different species), originally from the same slides, allowing for retrieval of this information by investigators who have only a portion of the original series.

Types. According to Rosen and DeBach (1979), the holotypes (in some cases lectotypes, neotypes, syntypes and allotypes) of the species of *Aphytis* described by either P. DeBach and D. Rosen (sometimes by P. DeBach alone or with other collaborators) were deposited in the collection of the Department (later the Division) of Biological Control at UCR, which is now part of UCRC. Most of these types are now on permanent loan to the USNM but a few are missing. Schauff (1985) provided a list of the *Aphytis* types belonging to 56 species sent to the USNM by P. DeBach and J. Hall on permanent loan. Some of H. Compere's types of *Aphytis,* also listed by Rosen and DeBach (1979) as being deposited in the Division of Biological Control at UCRC, in fact belonged to the USNM, as Compere (1955) wrote (p. 273): "The holotypes and allotypes of the species described as new in this study are to be deposited in the United States National Museum...". Specimen depositions were verified and corrected to match the original descriptions.

The holotypes and some paratypes of *A. bangalorensis* Rosen and DeBach (related to the *proclia* species group), *A. sankarani* Rosen and DeBach (related to the *lingnanensis* and *mytilaspidis* species groups), and *A. landii* Rosen and DeBach (related to the *chrysomphali* and *mytilaspidis* species groups) were supposed to be deposited in UCRC (Rosen and DeBach 1986) but are all missing (and neither they could be found in USNM). No paratypes of *A. limonus* (Rust) (Rust 1915) can be now found in UCRC. Consequently, these species are not included in this catalog.

MATERIAL AND SUMMARY OF RESULTS

This annotated catalog of the type material of *Aphytis* in UCRC provides information on 75 valid species. A total of 7,390 specimens were remounted from Hoyer's medium into Canada balsam (one specimen per slide), 7,292 of these were remounted, labeled, and databased, and 98 specimens were remounted and labeled but not databased (the types of South African species received by J.-W. Kim on loan from the South African National Collection of Insects, Plant Protection Research Institute, Pretoria, South Africa).

309 primary type specimens (i.e., holotypes, syntypes, and neotypes) of the following 51 nominal species of *Aphytis* were remounted and relabeled: *A. acrenulatus* DeBach and Rosen, *A. acutaspidis* Rosen and DeBach, *A. africanus* Quednau, *A. amazonensis* Rosen and DeBach, *A. anneckei* DeBach and Rosen, *A. anomalus* Compere, *A. antennalis* Rosen and DeBach, *A. australiensis* DeBach and Rosen, *A. cercinus* Compere, *A. citrinus* Compere, *A. cochereaui* DeBach and Rosen, *A. coheni* DeBach, *A. comperei* DeBach and Rosen, *A. dealbatus* Compere, *A. debachi* Azim, *A. desantisi* DeBach and Rosen, *A. equatorialis* Rosen and DeBach, *A. fabresi* DeBach and Rosen, *A. fisheri* DeBach, *A. gordoni* DeBach and Rosen, *A. griseus* Quednau, *A. holoxanthus* DeBach, *A. hyalinipennis* Rosen and DeBach, *A. ignotus* Compere, *A. japonicus* DeBach and Azim, *A. lepidosaphes* Compere, *A. longicaudus* Rosen and DeBach, *A. luteus* (Ratzeburg), *A. malayensis* DeBach and Rosen, *A. mandalayensis* Rosen and DeBach, *A. margaretae* DeBach and Rosen, *A. mazalae* DeBach and Rosen, *A. melinus* DeBach, *A. obscurus* DeBach and Rosen, *A. paramaculicornis* DeBach and Rosen, *A. perplexus* Rosen and DeBach, *A. philippinensis* DeBach and Rosen, *A. phoenicis* DeBach and Rosen, *A. pinnaspidis* Rosen and DeBach, *A. riyadhi* DeBach, *A. rolaspidis* DeBach and Rosen (replacement name for *A. flavus* Quednau), *A. roseni* DeBach and Gordh, *A.*

7

salvadorensis Rosen and DeBach, *A. sensorius* DeBach and Rosen, *A. setosus* DeBach and Rosen, *A. spiniferus* Compere and Annecke, *A. taylori* Quednau, *A. theae* (Cameron), *A. tucumani* Rosen and DeBach, *A. vandenboschi* DeBach and Rosen, *A. yasumatsui* Azim.

2,473 secondary type specimens (i.e., paratypes, allotypes, and paralectotypes) of numerous nominal species of *Aphytis* were remounted, as well as 4,608 non-type specimens representing most of species in UCRC.

Lectotypes are here designated for *A. africanus* Quednau, *A. coheni* DeBach, *A. cylindratus* Compere, *A. equatorialis* Rosen and DeBach, *A. fisheri* DeBach, *A. holoxanthus* DeBach, *A. immaculatus* Compere, *A. lepidosaphes* Compere, *A. lingnanensis* Compere, *A. melinus* DeBach, and *A. taylori* Quednau.

Depositories of the primary types of the *Aphytis* species treated in this catalog are summarized in Appendix 1.

Abbreviations for the depositories of specimens are as follows: AMUZ, Aligarh Muslim University, Aligarh, Uttar Pradesh, India; ANIC, Australian National Collection of Insects, Canberra, Australian Capital Territory, Australia; BMNH, the Natural History Museum, London, England, UK; CNCI, Canadian National Collection of Insects, Ottawa, Ontario, Canada; DEI, Deutsches Entomologisches Institut, Müncheberg, Germany; EMEC, Essig Museum of Entomology, University of California, Berkeley, California, USA; IMLA, Fundación e Instituto Miguel Lillo, San Miguel de Tucumán, Tucumán, Argentina; IZCAS, Institute of Zoology, Chinese Academy of Sciences, Beijing, China; MLPA, Museo de la Plata, La Plata, Buenos Aires, Argentina; MNMS, Museo Nacional de Ciencias Naturales, Madrid, Spain; SANC, South African National Collection of Insects, Plant Protection Research Institute, Pretoria, South Africa; TAMU, Entomology Department, Texas A&M University, College Station, Texas, USA; UCDC, Bohart Museum of Entomology, University of California, Davis, California, USA; UCRC, Entomology Research Museum, University of California, Riverside, California, USA; USNM, National

Museum of Natural History, Washington, D.C., USA; ZIN, Zoological Institute, Russian Academy of Sciences, St. Petersburg, Russia.

Species-group placements for the included species of *Aphytis* follow those given by Rosen and DeBach (1979).

The first author is responsible for the text of this catalog except for the species in the *lingnanensis* species group, which was prepared together by Jung-Wook Kim and Serguei V. Triapitsyn.

METHODS

REMOUNTING PROCEDURE

This procedure generally follows Platner et al. (1999) with a few modifications that are better suited for remounting *Aphytis* specimens. A significant problem in remounting is that the antennae may collapse after contacting Canada balsam. The following method prevented major collapsing in about 90% of the specimens. The success rate in remounting and avoiding possible problems largely depends on the initial quality of the specimens: the better their original condition, the fewer problems are encountered during the process of remounting.

The standard remounting procedure was as follows.

1) Coverslips of Hoyer's mounts almost always were sealed with Zut®, glyptal, or other compounds to reduce desiccation. This sealant is removed with the tip of a razor blade before processing.

2) The slide is placed in a Petri dish and soaked in distilled water for 60-72 hours. After soaking, the coverslip can be lifted free of the specimen(s).

3) Specimen(s) is/are transferred with a hooked probe to a ceramic depression plate containing 10% ethanol, then covered with a 6 mm coverslip.

4) Ethanol is with a pipette and replaced with 10% KOH for 30-40 min. (if the specimen is already clear enough) or up to 24 hours (if the specimen has not been cleared in the original slide). At this point, the quality of the original specimen can be improved somewhat. KOH not only is a clearing and softening agent but also reduces head and antennal collapse. Therefore, specimens must be treated with KOH even if they had previously been cleared in the Hoyer's medium. Because of

flattening from the previous mount it is very difficult to reposition specimens during the remounting process.

5) KOH is removed with a pipette and replaced with 10% ethanol for 30 min. The dehydration procedure is repeated with 20%, 40%, 60%, 80%, 95%, and twice with 100% ethanol.

6) Absolute ethanol is with a pipette and replaced with a 1:1 mixture of 100% ethanol and clove oil. The depression plate with specimens is placed into a partly opened container for at least 24 hours (and up to 2-3 weeks if necessary) to allow for slow, complete evaporation of the alcohol and any remaining water. Once evaporation is complete only clove oil will remain and the specimen is ready for mounting.

7) Using the balsam applicator, a small drop of fresh (liquid) Canada balsam (pre-mixed with fresh 15% clove oil) is placed in the center of a clean slide, which sits in a template to assist proper positioning of the specimen at the center of the slide. A dot of mountant ca. 3-4 mm in diameter is optimum, but this dot should not be mixed or spread significantly. The specimen is gently placed in the mountant by removing it from the clove oil with a hooked probe and submerging it, dorsum up, in the balsam drop. At this point the specimen can be oriented and some minor repositioning of body parts is possible. It is important to do it quickly, not allowing the mountant to dry after specimen placement. A dry 6 mm coverslip (precleaned with 80% ethanol and lens tissue) is then immediately placed flat on the specimen. This allows it to contact the balsam drop near its center and forces air bubbles out when pressure is applied with the forceps.

8) The slide is marked with the individual code to assure proper labeling, and a corresponding male (σ) or female ($\female$) symbol is marked on the slide. The slide is then labeled in such a way that the specimen's head is positioned upside down relative to the label; this is necessary to ensure that when the specimen is viewed using a compound microscope, it appears in the right position with its head up.

9) The slide is placed onto a slide warmer (50°C) for at least 7 days.

LABELING

The original slide (usually without specimens) is kept for historical purposes and is the main source of the labeling information. If necessary, labels from the original slide were glued back onto it (Fig. 11). Because many slides did not contain proper labeling data, when possible we complemented the original data with additional, essential information available either from Rosen and DeBach (1979), other relevant publications, or from the UCR quarantine records. On top of this original slide, a new label was attached with the bar code number(s) of the slides that contained the specimen(s) remounted from this original slide. Example: "UCRC ENT 004390-004417" (Fig. 11). The original slides are stored separately in plastic 100-slide boxes, arranged numerically (according to the bar code number).

The new slides were labeled as follows (Fig. 12): the right label contains locality, date, collector's name, and other relevant data such as host record(s), plant record(s), P. DeBach's catalog number, any notes or other codes. The left label indicates the species identification, type status (if any), sex indication, name of the person responsible for identification and the date (if available). At the bottom, the left label states "Remounted from Hoyer's into Canada balsam [year] by [either] M. Planoutene or V. Berezovskiy (UCR)."

Finally, a bar code with the database number is affixed to the slide between the identification label and the coverslip (Fig. 12).

DATABASING

All label and other relevant data (about each specimen remounted on the new slide were entered into the Biota (Windows-95) database. When available, geographical coordinates, host information, etc. was added to the original label data. This database was later transferred to FileMaker Pro format and made available online at http://sanders5.ucr.edu:8080/. The bar code number was entered automatically using the Intermec bar code reader (Model 1550), connected to a Twinhead Slimnote Power Book (TE-200TZ).

ANNOTATED CATALOG OF SPECIES

Aphytis acrenulatus DeBach and Rosen

(related to *proclia* species group)

Aphytis acrenulatus DeBach and Rosen 1976: 543; Rosen and DeBach 1979: 419-421.

TYPE MATERIAL. Before remounting, there were 24 paratype specimens in UCRC on 24 slides, all mounted in dried Hoyer's medium. In addition, the holotype female and allotype male, mounted individually on slides in Hoyer's, were received from USNM.

Holotype female. Original labels: 1. (in red ink) "NAME *Aphytis* ♀ *acrenulatus* N. SP. HOLOTYPE ♀ DET. DR 1973 COLL. J. R. Williams No. IID.8:1 Div. Biol. Cont. Univ. Calif"; 2. (in black ink) "LOC. Montagne Longue, Mauritius DATE III.1971 HOST *Aspidiella zingiberi*". After remounting, the holotype is labeled as follows: 1. "*Aphytis acrenulatus* ♀ DeBach and Rosen <u>HOLOTYPE</u> Det. D. Rosen 1973 Remounted from Hoyer's into Canada balsam 2000 By M. Planoutene (UCR)"; 2. "Mauritius, Montagne Longue, iii.1971, J. R. Williams. Host: *Aspidiella zingiberi* Mamet. DeBach code IID.8:1"; 3. (bar code) "UCRC ENT 004588". The holotype, deposited in USNM (on permanent loan from UCRC), is not completely cleared but otherwise is in fair condition except for the forewings which are slightly damaged apically; the head and both antennae are detached from the body but are mounted closely together under the same coverslip.

Paratypes. Allotype male [USNM], same label data as holotype; 20 females and 4 males, same data as holotype, deposited as follows: 1 female [ANIC]; 1 female [CNCI]; 1 female [EMEC]; 1 female, 1 male [TAMU]; 1 female [UCDC]; 1 female, 1 male [USNM]; 1 female [ZIN]; remainder of paratypes in UCRC.

COMMENTS. According to both DeBach and Rosen (1976) and Rosen and DeBach (1979), the type series of *A. acrenulatus* consisted of 23 females and 4 males, including the holotype and the allotype. Of those, we remounted 26 specimens, so 1 female paratype is unaccounted for; it may have been deposited earlier in some other collection.

Aphytis acutaspidis Rosen and DeBach
(*vittatus* species group)

Aphytis acutaspidis Rosen and DeBach 1979: 248-249.

TYPE MATERIAL. Before remounting, there were 3 paratype specimens in UCRC on 3 slides, all mounted in darkened Hoyer's medium. In addition, the holotype female, on a slide in Hoyer's, was received from USNM. It was mounted together with a pupa that was not part of the type series. This pupa was also remounted into Canada balsam on a separate slide.

Holotype female. Original labels: 1. (in blue and red ink) "NAME *Aphytis* ♀ [*costa-limai* ? - crossed-out] *acutaspidis* n. sp. HOLOTYPE ♀ DET. BY DeB 1964 IA.10:1 LOT No. 45 DEPT BIOL. CONN. UNIV CALIF"; 2. (in blue and black ink) "LOC. Rural University, Rio de J. State, Brazil DATED July 14 1962 HOST *Acutaspis albopicta* (Cockerell) DET BY L. Arg. 19 on coconut palm COLL. DeBach". After remounting, the holotype is labeled as follows: 1. "*Aphytis acutaspidis* ♀ Rosen and DeBach <u>HOLOTYPE</u> Det. DeBach 1964 Remounted from

Hoyer's into Canada balsam 2000 By M. Planoutene (UCR)"; 2. "Brazil, Estado do Rio de Janeiro, Rural University, 14.vii.1962 P. DeBach. Host: *Acutaspis albopicta* (Cockerell) on coconut palm (Det. by L. Arg.). No. on orig. slide "45" "*costa-limai*?" crossed-out DeBach code IA.10:1"; 3. (bar code) "UCRC ENT 004285". The holotype, deposited in USNM (on permanent loan from UCRC), is in good condition and mounted under one coverslip; one forewing, several leg segments, and the head are detached from the body; only one flagellum is present, it is detached from the scape and pedicel, both pairs of which are attached to the head.

Paratypes. 3 females, same data as holotype, are deposited as follows: 1 female [USNM]; the other two in UCRC. Two of the paratypes are lacking their heads.

Aphytis africanus Quednau

(*lingnanensis* species group)

Aphytis africanus Quednau 1964: 112-113; Rosen and DeBach 1979: 542-545.

TYPE MATERIAL. Before remounting, there were 30 syntype specimens in UCRC on 3 slides, all mounted in Hoyer's, and the medium was completely black (Fig. 8). Originally, there were 35 females and 31 males, all designated as syntypes (Quednau 1964). Rosen and DeBach (1979) examined 20 females and 14 males out of the original syntype series.

Lectotype female, here designated to avoid confusion about the status of the remounted type specimens of *A. africanus*. Original labels: 1. (red) "*Aphytis africanus* Quednau ♀ Syntypes W. Quednau det."; 2. "Red scale on oranges, South Africa, Transvaal, Letaba, 26.5.61 [followed in pencil] Quednau TYPE Box IVB.12:1 [DeBach's code]". The lectotype female was originally mounted on the same slide with 14 other syntypes. After remounting, the lectotype is labeled as follows: 1. "*Aphytis africanus* ♀ Quednau SYNTYPE Remounted from Hoyer's into Canada balsam 1999 By M. Planoutene (UCR)"; 2. "*Aphytis africanus* ♀ Quednau LECTOTYPE des. by J.-W. Kim and S. V. Triapitsyn 2002"; 3. "South Africa, Transvaal, Letaba, 26.v.1961. Host: *Aonidiella aurantii* (Maskell) on oranges. Note on orig. "Qued Type Box". DeBach Code IVB.12:1"; 4. (bar code) "UCRC ENT 001485". The lectotype, deposited in USNM (on permanent loan from UCRC), is in fair condition and complete.

Paralectotypes. 14 females (remounted from the same original slide as the lectotype) [1 in BMNH, 1 in CNCI, 1 in SANC, 1 in TAMU, 1 in USNM, remainder in UCRC]; 11 males [1 in SANC, 1 in USNM, remainder in UCRC] remounted (on to 10 slides) from a separate slide with the same label data as lectotype except for DeBach's code (IVB.12:2), and originally labeled as male syntypes; and also 4 females [UCRC] remounted individually from the original slide labeled: 1. (red) "*Aphytis africanus* Quednau ♂ ♀ Syntypes W. Quednau det."; 2. "Red scale on oranges, South Africa, Transvaal, Letaba, 1.10.62 [followed in pencil] Quednau TYPE Box IVB.12:3 [DeBach's code]".

NON-TYPE MATERIAL. Additionally, remounted were 376 non-type specimens of *A. africanus* as well as 18 specimens identified by D. Rosen in 1969 as *A. ?africanus*. 635 specimens identified as "*Aphytis africanus* or *melinus*" or "*Aphytis africanus* or *melinus* or *lingnanensis*" were also remounted [all in UCRC].

COMMENTS. Although initially confused with the biparental form of *A. chrysomphali* (Mercet), *A. africanus* was later proved to be reproductively isolated from both *A. chrysomphali* and *A. lingnanensis* Compere (Quednau 1964). P. DeBach noted that *A. africanus* is very difficult to distinguish from *A. lingnanensis* (Quednau 1964). *Aphytis africanus* is indeed morphologically similar to *A. lingnanensis* and also to *A. coheni* DeBach. The dark color of the posterior margin of the scutellum in *A. africanus* may be one of its distinguishing features, yet it cannot be used on its own. Both *A. coheni* and *A. africanus* have a dark posterior part of the scutellum and a short ovipositor (relative to the mesotibia). Also, in both these species the scutellar sensilla are distinctly close to the anterior scutellar setae. *Aphytis africanus* can be separated from *A. lingnanensis* based on the dark pupa, a short ovipositor and antennae, and a weak stippling of male abdominal sterna. Small and less elongated crenulae of *A. africanus* are also a very distinctive feature.

Aphytis amazonensis Rosen and DeBach

(related to *proclia* species group)

Aphytis amazonensis Rosen and DeBach 1979: 422-424.

TYPE MATERIAL. Before remounting, there was one paratype specimen in UCRC on a slide, mounted in Hoyer's, in fair condition, as well as two empty slides, similarly labeled, as follows: 1."*Aphytis* 2 n. spp. nr. *lingnanensis*, nr. *proclia* ?? Det. DeB. 1964 Lot # 48(2)"; 2."Acapa Terr., Brazil, July 25 1962 on wild jungle plant Coll. DeBach"; 3. ([written in green ink on the glass] "Original slide from which *braziliensis* n. sp. was remounted". In addition, the holotype female and the allotype male, mounted separately on slides in Hoyer's, were received from USNM.

Holotype female. Original labels: 1. (in black and red ink) "NAME *Aphytis* ♀ [*braziliensis* - crossed-out] *amazonensis* n. sp. HOLOTYPE ♀ DET. DR 1969 COLL. P. DeBach No. 48 Dept. Biol. Cont. Univ. Calif."; 2. (in black ink) "LOC. Acapa, Terr., Brazil DATE July 25 1962 HOST IID.11:1 DET 19 ON wild jungle plant". After remounting, the holotype is labeled as follows: 1. "*Aphytis amazonensis* ♀ Rosen and DeBach HOLOTYPE Det. D. Rosen 1969 Remounted from Hoyer's into Canada balsam 2000"; 2. "Brazil, Acapa Territory, 25.vii.1962, P. DeBach. Host: undetermined armored scale on wild jungle plant. Note original label says: *Aphytis* [*braziliensis* - crossed-out] DeBach code IID.11:1 No. 48"; 3. (bar code)

"UCRC ENT 004592". The holotype, deposited in USNM (on permanent loan from UCRC), is in fair condition although it is not perfectly cleared; unfortunately, the flagellum of one antenna is missing and the flagellum of the other is obscured by the head in such a way that it is hardly visible, actually worse than it was on the original slide, as seen in the photograph (Fig. 718, p. 458 in Rosen and DeBach 1979).

Paratypes. Allotype male [USNM] and 1 paratype male [UCRC], with the same label data as holotype, both in fair condition.

Aphytis anneckei DeBach and Rosen
(*chrysomphali* species group)

Aphytis anneckei DeBach and Rosen 1976: 544; Rosen and DeBach 1979: 598-601.

TYPE MATERIAL. Before remounting, there were 2 paratype specimens in UCRC on separate slides, mounted in Hoyer's. In addition, the holotype female and the allotype male, mounted on the same slide in Hoyer's, were received from USNM.

Holotype female. Original labels (in red ink and pencil): 1. "S&R 1858 ♀ ♂ *Aphytis* types [an illegible word crossed-out] n. sp. Ex. *Chrysomphalus* sp. on *Trichillia emetica* Pretorius Kop, S. Africa July 19, 1958 B. R. Bartlett Coll. S&R 1858"; 2. "NAME *Aphytis anneckei* n. sp. HOLOTYPE ♀ ALLOTYPE ♂ DET. DR, DB 1973 COLL. No. VIB:15:1 Div. Biol. Cont. Univ. Calif." After remounting, the holotype is labeled as follows: 1. "*Aphytis anneckei* ♀ HOLOTYPE DeBach and Rosen Det. D. Rosen and P. DeBach 1973 Remounted by V. Berezovskiy (UCR)"; 2. "South Africa, Pretorius Kop, 19.vii.1958, B.R. Bartlett. Host: *Chrysomphalus* sp. on *Trichilia emetica* S&R 1858 DeBach code VIB.15:1"; 3. (bar code) "UCRC ENT 007291". The holotype, deposited in USNM (on permanent loan from UCRC), is in fair condition; the head is detached from the body and the pedicel and flagellum of one antenna are missing. Note that Rosen and DeBach (1979) indicated a different collecting date for the holotype and the allotype: July 10, 1958.

Paratypes. Allotype male [USNM], with the same label data as holotype, in good condition (one forewing mounted under a separate coverslip); 2 females [UCRC]: Kenya, 31.iii.1969, D. Gerling, ex. *Paraselenaspidus madagascariensis* (Mamet) on *Dovyalis caffra* ("Shipment 7"), both in good condition.

NON-TYPE MATERIAL. Also remounted were 3 females from Ghana, determined by D. Rosen as *A. ?anneckei* [UCRC].

Aphytis anomalus Compere

(*vittatus* species group)

Aphytis anomalus Compere 1955: 286; Rosen and DeBach 1979: 275-276.

TYPE MATERIAL. The holotype of *A. anomalus* was received on loan from USNM, where Compere (1955) intended to deposit the holotypes and allotypes of his *Aphytis* species. Originally, according to Compere (1955), it was mounted on a slide in Canada balsam, but later remounted in Hoyer's (Rosen and DeBach 1979), who erroneously indicated UCRC as its depository.

Holotype female. Original labels: 1. (red) "Holotype"; 2. "*Aphytis anomalus* Compere HOLOTYPE Captured on *Cassia imperialis*, Bom Retiro, Estado Rio de Janeiro, Brasil, Sept. 8, 1934 H. Compere, Coll. IB.7:1 *Marietta* [in pencil]". After remounting, the holotype is labeled as follows: 1. "*Aphytis anomalus* ♀ Compere HOLOTYPE Remounted from Hoyer's into Canada balsam 2000 By M. Planoutene (UCR)"; 2. "Brazil, Estado Rio de Janeiro, Bom Retiro, 8.ix.1934, H. Compere. Host: *Cassia imperialis* "Rosen and DeBach 1979 remounted in Hoyer's medium" DeBach code IB.7:1"; 3. (bar code) "UCRC ENT 004357". The holotype, deposited in

USNM, is in good condition and mounted under one coverslip; one forewing and both hind wings are detached from the body.

NON-TYPE MATERIAL. Also remounted were 4 specimens of *A. anomalus* from Belo Horizonte, Minas Gerais, Brazil, out of 7 such specimens mentioned by Rosen and DeBach (1979) [UCRC].

Aphytis antennalis Rosen and DeBach

(related to *chilensis* species group)

Aphytis antennalis Rosen and DeBach 1979: 357-359.

TYPE MATERIAL. The holotype female, mounted on a slide in Hoyer's, was received from USNM.

Holotype female. Original labels: 1. (in black and red ink) "NAME *Aphytis* [*flandersi* (a manuscript name) - crossed-out] *antennalis* n. sp. HOLOTYPE ♀ DET. D. Rosen 1968 COLL. S. E. Flanders No. S&R 1227-3 Dept. Biol. Cont. Univ. Calif."; 2. (in black ink) "LOC. Taipo Hong Kong DATE III-26, 1954 HOST ? ID.19:1 DET 19 ON citrus (*Aphytis* B [in circle]) Hoyer mount, Raymond 1962". After remounting, the holotype is labeled as follows: 1. "*Aphytis antennalis* ♀ Rosen and DeBach HOLOTYPE Det. D. Rosen 1968 Remounted from Hoyer's into Canada balsam 2000 By M. Planoutene (UCR)"; 2. "Hong Kong, Taipo, 26.iii.1954, S. E. Flanders. Host: unknown, on citrus. Note on original slide: "(*Aphytis* B [in circle]; Hoyer mount, Raymond 1962" DeBach code ID.19:1 S&R #1227-3"; 3. (bar code) "UCRC ENT 004316". The holotype, deposited in USNM (on permanent loan from UCRC), is in rather poor condition although it is well cleared; it is broken in two large parts, and the wings are damaged.

NON-TYPE MATERIAL. Also remounted from the same slide with the holotype was one female not mentioned by Rosen and DeBach (1979) [UCRC]. It may be conspecific with the holotype but is in a very bad condition and thus difficult to identify.

Aphytis aonidiae (Mercet)

(*mytilaspidis* species group)

Aphelinus aonidiae Mercet 1911: 511-514.

Aphytis citrinus Compere 1955: 312-313.

Aphytis aonidiae (Mercet): Rosen and DeBach 1979: 476-483 (list of synonyms and redescription).

TYPE MATERIAL. Before remounting, there were the following type specimens of *A. citrinus* in UCRC on slides: the lectotype female and 13 paralectotypes (labeled only as "cotypes"), all mounted in a dried Hoyer's medium, as well as 4 Canada balsam slides with numerous paralectotypes (all also labeled as "cotypes").

Lectotype female, designated by Rosen and DeBach (1979). Original labels: 1. (in red ink) "*Aphytis citrinus* Compere LECTOTYPE ♀ from cotype series DET. DR 1973 COLL. O. Hemphill NO. Hoyer remount 1969 [an illegible word crossed out] Div. Biol. Cont. Univ. Calif."; 2. (in black ink) "LOC. Visalia, Calif. DATE XII-20, 1948 HOST *Aonidiella citrina* IIIB.12:1 DET 19 ON orange (remounted in Hoyer)". After remounting, the lectotype is labeled as follows: 1. "*Aphytis citrinus* ♀ Compere LECTOTYPE (from cotype series) Det. D. Rosen 1973 Remounted from Hoyer's into Canada balsam 2000 By M. Planoutene (UCR)"; 2. "U.S.A., CA., Visalia, 20.xii.1948, O. Hemphill. Host: *Aonidiella citrina* on orange. Note on orig.

slide "remounted in Hoyer 1969", DeBach code IIIB:12:1. Note: "synonym of *A. aonidiae*"; 3. (bar code) "UCRC ENT 004942". Note that the slide number (DeBach code) of the lectotype does not agree with the one published by Rosen and DeBach (1979) (IIA2:9). The lectotype, deposited in USNM (on permanent loan from UCRC), is in good condition except one hind wing is missing.

Paralectotypes [UCRC], designated by Rosen and DeBach (1979), but before remounting all labeled as "cotypes": 13 females on separate slides, all remounted from Hoyer's, with the same label data as holotype except for DeBach codes from IIIB.12:6 to IIIB.12:18 and various notes on some of the slides pertinent to the condition of the specimens, such as "1 ant.[enna]", etc., and also 4 original Canada balsam slides made by H. Compere, labeled as follows: "*Aphytis citrinus* Compere COTYPES Ex. *Aonidiella citrina* on Orange from Oscar Hemphill, Visalia, Tulare Co., Calif., Dec. 20, 1948 Sent to S. E. Flanders". According to a note on one of the slides, some of the specimens from it were remounted into Hoyer's (slides # IIIB.12:1, 6-18). Neither Compere (1955) nor Rosen and DeBach (1979) counted the specimens on apparently 4 original Canada balsam slides that comprised the syntype series; the remaining 4 slides now have the following number of specimens: slide # 2 (DeBach code IIIB.12:2) - 9 whole females under one coverslip and about 10 fragmented females under 3 coverslips; slide # 3 (DeBach code IIIB.12:3) – 9 to 12 fragmented females under 2 coverslips (the lectotype and the paralectotypes were remounted into Hoyer's from this slide); slide # 4 (DeBach code IIIB.12:4) - 9 whole females under one coverslip and about 12 fragmented females under 2 coverslips; slide # 5 (DeBach code IIIB.12:5) - 14 whole females under one coverslip and about 13 fragmented females under the other coverslips. All paralectotypes including the ones on the original Canada balsam slides are now clearly marked as such.

NON-TYPE MATERIAL. Also remounted were 104 non-type specimens of *A. aonidiae* (some determined as *A. citrinus*) as well as 3 specimens determined as "*A. ?aonidiae, A. ?citrinus, A.* nr. *aonidiae* or *A.* nr. *citrinus*, from various countries [UCRC].

Aphytis australiensis DeBach and Rosen

(*vittatus* species group)

Aphytis australiensis DeBach and Rosen 1976: 542; Rosen and DeBach 1979: 272-274.

TYPE MATERIAL. Before remounting, there were 6 paratype specimens in UCRC on slides, 5 mounted in Hoyer's and one in Canada balsam. Also present in UCRC is a Canada balsam slide # IB.5:8 with the fragments of a non-type female, which has the same label data as the type series but was not mentioned either by DeBach and Rosen (1976) or Rosen and DeBach (1979). In addition, the holotype female, mounted on a slide in Hoyer's, was received from USNM.

Holotype female. Original labels: 1. (in red and black ink) "TYPE ♀ NAME *Aphytis australiensis* n. sp. HOLOTYPE DET. DR 1969 COLL. E. M. Ehrhorn NO IB.5:1 TYPE Dept. Biol. Cont. Univ. Calif."; 2. (in black ink) "LOC. Waroongus [a misspelling of Wahroonga], NSW, Australia DATE VIII-26, 1923 HOST *Chionaspis* sp. on mistletoe DET 19 ON *Eucalyptus* remount: Hoyer, from point, Tim' collection". After remounting, the holotype is labeled as follows: 1. "*Aphytis australiensis* ♀ DeBach and Rosen HOLOTYPE Det. D. Rosen 1969 Remounted from Hoyer's into Canada balsam 2000 By M. Planoutene (UCR)"; 2. "Australia, New South Wales, Waroongus, 26.viii.1923, E. M. Ehrhorn. Host: *Chionaspis* sp. on mistletoe on *Eucalyptus* "Remount: Hoyer from point. Timberlake collection". DeBach code IB:5:1"; 3. (bar code) "UCRC ENT 004356". The holotype, deposited in USNM (on permanent loan from UCRC), is in fair condition but is undercleared; tips of both forewings and one hind wing are missing.

Paratypes: 1 female [ANIC], 1 female [CNCI], 3 females [UCRC], 1 female [USNM] with the same label data as holotype except for DeBach codes IB.5:2 to IB.5:7.

Aphytis bedfordi Rosen and DeBach
(unassigned to species group)

Aphytis bedfordi Rosen and DeBach 1979: 692-694.

TYPE MATERIAL. Rosen and DeBach (1979) indicated that the only two known specimens of this species, the holotype and the paratype, both females, were deposited in UCRC. Later, both specimens were returned to SANC, apparently because they had been loaned from that collection, not just given to, D. Rosen and P. DeBach.

Aphytis capensis DeBach and Rosen
(*mytilaspidis* species group)

Aphytis capensis DeBach and Rosen 1976: 544; Rosen and DeBach 1979: 499-502.

TYPE MATERIAL. Before remounting, there were the following type specimens of *A. capensis* in UCRC on slides: the allotype male and 16 paratypes mounted individually on slides in dried Hoyer's, and 5 female paratypes mounted on 2 slides (2 and 3) in Canada balsam. In addition, a Canada balsam slide containing the holotype female and a paratype female, was received from USNM. This species was described from 15 female and 13 male specimens (DeBach and Rosen 1976); we have accounted for all but one female and 1 male paratypes on slides (these were given to SANC long ago); the latter most likely are still mounted in Hoyer's.

Holotype female (mounted on the same slide with one female paratype). Original labels: 1. (in red and black ink) "NAME *Aphytis capensis* DeB. and R. HOLOTYPE ♀ PARATYPE ♀ (Dissec.) DET. DeB and DR 1973 [COLL. - crossed out] in balsam NO [IA23:1 - crossed out] IIIC.20:1 Div. Biol. Cont. Univ. Calif."; 2. (in black ink and pencil) "♀ Boiled KOH. X *Aphytis* [*mytil.* - crossed out] ♀ Ex. *Chionaspis margaritae* Camp's Bay, C.P. So. Africa. July 13, 1925 E. W. Rust, Coll.". The holotype, deposited in USNM (on permanent loan from UCRC), is intact and in very good condition; the paratype female on the same slide is dissected (head, one antenna and a forewing are detached).

Paratypes. Allotype male [USNM] on slide labeled (new label): "South Africa, Cape Province, Camp's Bay, 10.vi.1925, E. W. Rust. Host: *Dentachionaspis margaritae* (Brain) DeBach code IIIC.20:2". Also the above female, mounted with the holotype; 6 females and 12 males on individual slides, same location, collector, and host as allotype except the dates are different: 25.iv.1923 - 3 males [ANIC, CNCI, TAMU], 26.iv.1923 - 2 females [ANIC, CNCI], 19.vii.1923 - 1 male, vii.1923 - 2 males [UCRC], 14.x.1923 - 1 female [TAMU], 28.i.1924 - 1 female, 12.vi.1925 - 1 female, 4.vii.1925 - 1 male, 18.vii.1925 - 2 males, 19.vii.1925 - 1 female, 2 males, no date, - 1 male [UCRC]; 2 females under one coverslip on slide, labeled: "♀ Boiled KOH. X *Aphytis* ♀ Ex. *Chionaspis margaritae* Camp's Bay, C.P. So. Africa. July 13, 1925 Rust, Coll. [Feb. 1950 good - in pencil"; 3 females under two coverslips on slide, labeled: "IIIC.21:3 Found in vial Rust's Cape Town material - designated X by Rust *Aphytis* not *mytilaspidis*" [both slides in UCRC].

Aphytis cercinus Compere

(*chilensis* species group)

Aphytis cercinus Compere 1955: 302; Rosen and DeBach 1979: 346-349.

TYPE MATERIAL. The holotype female, mounted in Hoyer's on a slide together with 5 female paratypes, was received on loan from USNM. Originally, according to Compere (1955), this species was described from 12 females including the holotype; it seems that the entire type series was originally mounted on at least (and likely) two slides, so the remaining 6 paratype females, not seen by Rosen and DeBach (1979), should be in BMNH where Compere (1955) intended to deposit some secondary types of his *Aphytis* species.

Holotype female. Original labels: 1. "*Aphytis cerci* Compere and Gahan 1 ♀ on left [right - crossed out] side Holotype 5 ♀ adjacent paratypes No. 4 of Gahan's letter"; 2. (red) "*cercinus* Compere [+ Gahan - crossed out]"; 3. "Rust's No. A15 Ex. *Aspidiotus* on mistletoe 1C.21:1 Durban, Natal, Dec. 18, 1925. E. W. Rust, Coll.". After remounting, the holotype is labeled as follows: 1. "*Aphytis cercinus* ♀ Compere HOLOTYPE Remounted from Hoyer's into Canada balsam 2000 By M. Planoutene (UCR)"; 2. "South Africa, Natal, Durban, 18.xii.1925, E. W. Rust. Host: *Aspidiotus* on mistletoe, Rust's No. A15. Note on orig. slide "No. 4 of Gahan's letter" DeBach code IC.21:1"; 3. (bar code) "UCRC ENT 004304". The holotype, deposited in USNM, is in poor condition and mounted under one coverslip; one forewing, a flagellum, and one middle leg (except coxa and trochanter) are detached from the body; the remaining flagellum of the other antenna, a forewing and both hind wings, and also tibia and tarsus of the other middle leg are missing.

Paratypes. 5 females on separate slides, some in better condition than the others but specimens more or less complete, with the same label data (new label) as holotype, deposited as follows: 1 female [ANIC], 1 female [CNCI], 2 females [UCRC], 1 female [USNM].

NON-TYPE MATERIAL. Also remounted were 15 non-type specimens of *A. cercinus* from South Africa [UCRC]; some of the specimens mentioned by Rosen and DeBach (1979) on 36 slides were long ago returned to SANC and thus were not available for remounting.

Aphytis chilensis Howard
(*chilensis* species group)

Aphytis chilensis Howard 1900: 168; Rosen and DeBach 1979: 349-354.
Aphelinus capitis Rust 1915: 73-74.
Aphytis riadi Delucchi 1964: 136-139.

TYPE MATERIAL. Before remounting, there was only one paratype female specimen of *A. riadi* Delucchi, a synonym of *A. chilensis*, in UCRC, mounted in dried Hoyer's medium. After remounting, this specimen, mentioned by Rosen and DeBach (1979), was labeled as follows: 1. "*Aphytis riadi* ♀ Delucchi <u>PARATYPE</u> Remounted from Hoyer's into Canada balsam 1999 By M. Planoutene (UCR)"; 2. "Lebanon, Fanar, 9.v.1964. Host: *P. [Parlatoria] pergandii* [Comstock] DeBach code ID.1:8"; 3. (bar code) "UCRC ENT 001607". The specimen is in good condition.

In addition, two specimens of *Aphelinus capitis* Rust, another synonym of *A. chilensis*, marked as "cotypes" and mounted on separate slides in Canada balsam, were received from USNM. These are labeled as follows: "No. 1 [in pencil] *Aphelinus capitis.* Rust cotype [in red ink] ID.1:2 [DeBach code, in pencil] Ex *Aspidiotus camelliae* Sign. on *Hedera helix* (Ivy). 14647 B May 7, 1912. Santa

Barbara, Cal. P. H. Timberlake". The slide is numbered "2" by a black marker on the glass, apparently by D. Rosen or P. DeBach. The second slide of *Aphelinus capitis* (numbered "1") [USNM], which besides one female specimen also has one female head of another specimen, is labeled as: "No. 1 [in pencil] ID.1:1 [DeBach code, in pencil] *Aphelinus capitis.* Rust cotype [in red ink] Ex. *A. camelliae* Sign. on *Hedera helix* 14647B May 7, 1912. Santa Barbara, Cal. P. H. Timberlake". Originally, according to Rust (1915), this species was described from 28 female specimens from California, and he also unambiguously noted (p. 74) the following: "Type on slide labeled: *Aphelinus capitis*. Ex *Aspidiotus camelliae* Sign. on *Hedera helix* (Ivy). 14647 B May 7, 1912. Santa Barbara, Cal. P. H. Timberlake". Therefore, the type specimen mentioned by Rust (1915) must be considered holotype; indeed, it can be found in USNM (labeled in red ink as "type") under type number 40221. Thus, the above-mentioned two females are in fact paratypes; Gahan (1924) and Compere (1955) used the term "cotype" in the same sense as the term "paratype" is used now. At least 10 other female paratypes on slides, out of 27 specimens listed by Rust (1915), are in UCRC; these had no type designation labels and were labeled properly.

NON-TYPE MATERIAL. Also remounted were 310 non-type specimens of *A. chilensis* from various countries [UCRC].

Aphytis chrysomphali (Mercet)

(*chrysomphali* species group)

Aphelinus chrysomphali Mercet 1912a: 135-140; Rosen and DeBach 1979: 593-598.
Aphelinus quaylei Rust 1915: 75-76.

TYPE MATERIAL. The holotype of *Aphelinus quaylei* Rust, a synonym of *A. chrysomphali*, is in USNM under type number 40222. Indeed, Rust (1915) explicitly wrote (p. 76) "Type on slide labeled: 192°3b.; *Aphelinus quaylei* ex. *Pseudaonidia articulatus* on *Ficus nitidis*. Lima, Peru. January 31, 1914.—E. W. Rust". Therefore, another specimen in USNM is in fact a paratype; it has the same label data as the holotype, with an additional label "This must be a Cotype HC. Dec. 1951 VIA1:9 [DeBach code, in pencil]". At least 8 other female paratypes on individual slides, out of "many female specimens" from Peru and California mentioned by Rust (1915), are in UCRC; these had no type designation labels and were mistakenly mentioned by Rosen and DeBach (1979) as "syntypes". We labeled them properly as paratypes. Two of those (with the same label data as the holotype) were remounted from Hoyer's into Canada balsam. The label data of the remaining paratypes (4 in Hoyer's, 2 in Canada balsam) are as follows: 2 females - Peru, Piura, Mallares, 22.II.1912, E. W. Rust (number "14°3b"); 1 female - USA, California, Orange Co., Avondale, 23.IX.1911, P. H. Timberlake ("Ex *Chrysomphalus aurantii citrinus* 14527D"); 1 female - California, Santa Barbara Co., Carpinteria, 9.XI.1911, P. H. Timberlake ("Ex *Chrysomphalus aurantii citrinus* on orange 14576E"); 1 female - California, Los Angeles Co., Whittier, 22.XI.1910, P. H. Timberlake ("Ex *Chrysomphalus aurantii* on orange tree 14503A"); 1 female – Whittier, 11.IX.1911, P. H. Timberlake ("Ex *Chrysomphalus aurantii* 14503C").

NON-TYPE MATERIAL. Also remounted were 3 non-type specimens of *A. chrysomphali* from Spain and Tahiti and 2 specimens from Japan (identified as *A. ?chrysomphali*) [UCRC].

Aphytis cochereaui DeBach and Rosen

(*vittatus* species group)

Aphytis cochereaui DeBach and Rosen 1976: 541-542; Rosen and DeBach 1979: 251-256.

TYPE MATERIAL. Before remounting, there were 47 paratype specimens in UCRC mounted on 16 slides in dried or darkened Hoyer's, and 5 female paratypes mounted on 2 slides (2 and 3) in Canada balsam. In addition, the holotype female, the allotype male, and 7 paratypes were received from USNM; these also were mounted in dried Hoyer's.

Holotype female (originally mounted on the same slide with the remains of the host; those were remounted onto a separate slide). Original labels: 1. (in red and black ink) "NAME *Aphytis cochereaui*, n. sp. HOLOTYPE ♀ DET. DeBach 19 DR COLL. DeBach and Cochereau Dept. Biol. Cont. Univ. Calif."; 2. (in black and blue ink and pencil) "LOC. Sarramea, New Caledonia DATE IV-24 1968 HOST *Lepidosaphes beckii* DET. DeBach 1968 ON *Murrayea exotica* IA.14:1". After remounting, the holotype is labeled as follows: 1. "*Aphytis cochereaui* ♀ DeBach and Rosen HOLOTYPE Det. D. Rosen Remounted from Hoyer's into Canada balsam 2000 By M. Planoutene (UCR)"; 2. "New Caledonia, Sarramea, 24.iv.1968, DeBach and Cochereau. Host: *Lepidosaphes beckii* (Newman) on *Murrayea exotica* (Det. by DeBach) DeBach code IA.14:1"; 3. (bar code) "UCRC ENT 004287". The holotype, deposited in USNM (on permanent loan from UCRC), is in a fair condition, with the head detached from the body; one hind wing is missing.

Paratypes. Allotype male on slide labeled (new label): "New Caledonia, Noumea, x.1967, Cochereau. Host: a mixture of armored scale insects on *Citrus nobilis* No. on orig. slide "3" DeBach code IA.14:2" [USNM]. Also remounted were 59 other paratype specimens (34 females and 25 males [all in UCRC except for 1 female, 1 male in USNM]) from New Caledonia, with various label data exactly as indicated for the paratypes of *A. cochereaui* by Rosen and DeBach (1979), not listed here to avoid unnecessary repetition.

NON-TYPE MATERIAL. Remounted were 27 non-type specimens of *A. cochereaui* from New Caledonia [UCRC].

Aphytis coheni DeBach

(*lingnanensis* species group)

Aphytis coheni DeBach 1960: 705; Rosen and DeBach 1979: 539-542.

TYPE MATERIAL. Before remounting, there were numerous syntypes, labeled as "paralectotypes", in UCRC on 3 slides in a dark Hoyer's medium. We also received the "lectotype" female and a "paralectotype" male, mounted on the same slide with two other specimens of *Aphytis* from USNM. However, this "lectotype" female was designated by Rosen and DeBach (1979) invalidly because it was not part of the syntype series designated in the original description by DeBach (1960). The slide with this "lectotype" also contained one male "paralectotype" of *A. coheni* (Rosen and DeBach 1979, also an invalid designation) and two other specimens, one of which was identified by D. Rosen as a male *A. holoxanthus* DeBach and the other was a female *A. coheni* not mentioned by Rosen and DeBach (1979). After remounting, the identities of these specimens were verified by J.-W. Kim and labeled accordingly. The new lectotype was then selected from the original syntype series of *A. coheni* to clarify the existing confusion about the type specimens of this species.

Lectotype female, here designated. Original labels: 1. (bordered in red ink) "*Aphytis coheni* (DeBach Syntypes) reared on Oleander scale on lemon"; 2. (in blue ink) "Israel, reared in insectary, CES, Riverside, Calif., May 1960, Host: *A. aurantii* (Israel), on citrus (Israel), Nadell [sic] (Israel)"; 3. "Paralectotypes, R. S. 91". After remounting, the lectotype is labeled as follows: 1. " *Aphytis coheni* ♀ DeBach LECTOTYPE des. By J.-W. Kim and S. V. Triapitsyn Remounted from Hoyer's into Canada balsam 2000 By V. Berezovskiy (UCR)"; 2. "U.S.A., Ca., Riverside Co., UCR Insectary culture, v. 1960. Host: *Aspidiotus nerii* Bouché on lemon. Note on orig. slide "originally *A. aurantii* California red scale in Israel" DeBach code IVB.1:2"; 3. (bar code) "UCRC ENT 004492". The lectotype, deposited in USNM (on permanent loan from UCRC), is in fair condition, with head, flagellum of the right antenna, and one pair of wings detached from the body.

Paralectotypes. 17 females, 3 males, and 8 pupae [UCRC] were remounted from the same slide with the lectotype. Additional 43 females, 12 males, and 20 pupae [1 female and 1 male in BMNH, 1 female and 1 male in CNCI, 1 female and 1 male in TAMU, 1 female and 1 male in USNM, 1 female and 1 male in ZIN, remainder in UCRC] were remounted from other two syntype slides mentioned by DeBach (1960), with the identical original labels and different DeBach's numbers (DeBach codes IVB.1:2 and IVB.1:3, respectively).

NON-TYPE MATERIAL. Also remounted were 14 specimens of *A. coheni* from Israel and 4 pupae identified as "*A. coheni*–Khunti'' [UCRC].

Aphytis comperei DeBach and Rosen
(*proclia* species group)

Aphytis comperei DeBach and Rosen 1976: 543; Rosen and DeBach 1979: 394-397.

TYPE MATERIAL. Before remounting, there were about 130 paratype specimens in UCRC mounted on 14 slides in dried or darkened Hoyer's. In addition, the holotype female of *A. comperei* was received from USNM; it also was mounted in a partially dried Hoyer's medium.

Holotype female. Original labels: 1. (in red and black ink) "NAME *Aphytis* n. sp. nr. *hispanicus comperei* n. sp. HOLOTYPE ♀ DET. DR 1970 COLL. W. Hart NO. 1 IIB.16:1 Dept. Biol. Cont. Univ. Calif."; 2. (in black ink) "LOC. Texas McAllen DATE Aug 1968 HOST *Aonidiella aurantii* DET. 19 ON ?citrus [HOST SCALE = ???]". After remounting, the holotype is labeled as follows: 1. "*Aphytis comperei* ♀ DeBach and Rosen <u>HOLOTYPE</u> Det. D. Rosen 1970 Remounted from Hoyer's into

Canada balsam 2000 By M. Planoutene (UCR)"; 2. "USA: TX, McAllen, viii.1968, W. Hart. Host: *Aonidiella aurantii* on ?citrus. Note on orig. slide "n. sp. nr. *hispanicus*;" DeBach code IIB.16:1"; 3. (bar code) "UCRC ENT 004795". The holotype, deposited in USNM (on permanent loan from UCRC), is in fair condition but the body is somewhat shriveled, one forewing is detached from the body and its tip is folded.

Paratypes. There is a huge problem with the actual number of the paratypes of *A. comperei*. DeBach and Rosen (1976) only specified the locality and some other label data for the holotype of this species, also stating that their new species was described from numerous female specimens. Later, Rosen and DeBach (1979) redescribed it from the entire type series that included the holotype and other, numerous specimens for which label data were indicated, so these should be considered paratypes. Indeed, such slides in UCRC do bear paratype labels. However on many of them the paratype specimens of *A. comperei* were mixed with some non-type specimens of *A. hispanicus* (Fig. 10), but the paratypes of the former usually were not marked as such. We find those two species very closely related, if not conspecific, because morphologically intermediate forms between them do occur. Therefore, separating some paratypes of *A. comperei* from non-type specimens of *A. hispanicus* was impossible either before or after remounting. After remounting, we had little choice but to label such specimens simply as "either a paratype of *A. comperei* or *A. hispanicus*", as follows: 26 females and 6 pupae (the pupae were not part of the type series), labeled: "USA, Texas, 30.vi.1965. Host: *Parlatoria pergandii* on citrus. DeBach Code IIB.16:2 S&R # 65-49" as well as 11 females on slides, labeled: "Mexico, Sinaloa, Los Mochis, 17.i.1967, P. DeBach. Host: *Parlatoria pergandii* on sour orange. DeBach Code IIB.16:5" (5 female specimens) or "DeBach Code IIB.16:6" (6 female specimens) [all in UCRC]. Altogether, 120 likely paratype females of *A. comperei* were remounted individually on slides, out of the 147 females belonging to the type series as mentioned by Rosen and DeBach (1979), who also provided their label data, omitted here to avoid repetition (note that the locality

Denbigh Kraal, Jamaica, was misspelled by them as "Denbign Kraal). A paratype female (on a separate slide, in Hoyer's) was sent earlier to SANC (according to the note left in the collection). 23 female paratypes in UCRC (mounted on one slide in Hoyer's together with 8 females of *A. hispanicus*) were left in the original mounting medium (label data: "Joalmi, Clarendon, Jamaica, 28.ii.1968, coll. L. W. van Whervin, ex. "purple, green (soft) scale").

NON-TYPE MATERIAL. Also remounted were 4 non-type specimens of *A*. nr. *comperei* from Hong Kong (China) [UCRC].

Aphytis confusus DeBach and Rosen

(*proclia* species group)

Aphytis confusus DeBach and Rosen 1976: 543; Rosen and DeBach 1979: 402-405.

TYPE MATERIAL. Before remounting, there were 4 paratype specimens (2 females and 2 males) in UCRC on 2 slides, all mounted in Hoyer's; one of them was completely dark. The rest of the type series (originally mounted on 6 slides), including the holotype female, the allotype male, and 9 paratype females and 1 paratype male, were listed as being kept at UCRC by Rosen and DeBach (1979) but later were returned to SANC (Schauff 1985); they most probably are still mounted in Hoyer's. The notes in UCRC indicate that the holotype female was mounted on the same slide with 2 female paratypes (DeBach code of the slide is IIC.1:1); the allotype male was mounted on the same slide with one female and 1 male paratypes (DeBach code of the slide is IIC.1:2), one female paratype was mounted individually on a slide (DeBach code is IIC.1:4), and 5 remaining female paratypes were mounted on the same slide (DeBach code of the slide is IIC.1:6). The label data for the holotype, allotype, and all the paratype specimens returned to SANC were the same, as indicated by Rosen and DeBach (1979) as well as in the original description.

Paratypes. Remounted were 1 female and 1 male from the slide with the DeBach code IIC.1:5 (same data as holotype), now labeled: "South Africa, Pretoria, Fountains, viii.1961, D. P. Annecke. Host: *Ledaspis distincta* (Leonardi) on *Protea caffra* DeBach code IIC.1:5 S&R T-576" [UCRC]; also 1 female and 1 male from the slide with DeBach code IIC.1:3, now labeled: "South Africa, Natal, Ngwavuma, ix.1961, D. P. Annecke. Host: *Africaspis chionaspiformis* (Newstead) on *Melanaspis corticosa* (Brain) DeBach code IIC.1:3" [UCRC].

Aphytis costalimai (Gomes)

(*chilensis* species group)

Marietta costa-limai Gomes 1942: 23-25, est.1.

Aphytis spiniferus Compere and Annecke 1961: 27-28.

Aphytis costalimai (Gomes): Rosen and DeBach 1979: 244-248.

TYPE MATERIAL. Before remounting, there was 1 paratype female specimen of *A. costalimai* in UCRC, partially mounted in Hoyer's, and partially in Canada balsam on the same slide. The original slide, which still has at least the flagellum of one antenna remaining in Canada balsam, is labeled as follows: 1."No. 016 *Marietta aonidiphaga* n.sp. ♀ E. do Rio, 15/1/40 J. Gomes, col. J. Gomes, det."; 2."ESC. NAC. DE AGRONOMIA Paratype [apparently added in blue ink later by P. DeBach] Preparação 80 Divisão 30' IA.7:2 [DeBach code in pencil] Caixa 1 *Marietta costa-limai* [apparently added in blue ink later by P. DeBach]"; 3. (on the back of slide, in blue ink) "DeBach 1962 Note: The manuscript name "*aonidiphaga*" was not used, this species was described as *costa-limai*. The change was made by Gomes with the reference card for the paratype slides, but not on the slides. Later Compere and Annecke (1961) called it *Aphytis spiniferus*". After remounting of the Hoyer's part, this specimen, mentioned by Rosen and DeBach (1979), was labeled as follows (data label): 1. "Brazil, Rio de Janeiro, 34 km on Rio de Janeiro - São Paulo Highway, 15.i.1940, J. G. Gomes. Host: *Chrysomphalus aonidum* on citrus. Original specific name "*Marietta aonidiphaga*", but species name was erased by pencil. No. on orig. slide "016" DeBach code IA.7:2". The remounted body of the specimen is in a very poor condition, with many parts missing including the wings.

In addition, most of the type series of *A. spiniferus* Compere and Annecke was remounted. This species was described originally from 15 females, the holotype and paratypes, and out of these we remounted the holotype, which was received on loan from USNM, and 3 paratypes that were in UCRC but were not marked as paratypes. We added appropriate paratype labels to these specimens, 6 of which (on 2 slides) were not remounted (because they are too fragmented) and remain in Hoyer's. Also in UCRC there are 5 females on individual points, bearing yellow paratype labels, most of them originally not sufficiently labeled individually, but collectively labeled as follows: 1. "S and R 1805"; 2. "*Marietta spiniferus* n. sp. H. Compere" or "*Marietta spiniferus* Comp."; 3. "Captured by beating citrus trees"; 4. "Limeira, Brazil", 5. "3/21/58 S. E. Flanders, Coll."; we added the appropriate data and identification labels to each of them.

Holotype female of *A. spiniferus*. Original labels: 1. (in black ink and pencil) "NAME *Aphytis spiniferus* DET. 19 COLL. NO. IA.7:1 Div. Biol. Cont. Univ. Calif."; 2. (in red and blue ink) HOLOTYPE S&R 1805 II-4 Limeira, Brazil 3-21-58 S. E. Flanders *Aphytis spiniferus* Ex. *Chrysomphalus on citrus*". After remounting, the holotype is labeled as follows: 1. "*Aphytis spiniferus* ♀ Compere and Annecke HOLOTYPE Remounted from Hoyer's into Canada balsam 2000 By M. Planoutene (UCR)"; 2. "Brazil, Limeira, 21.iii.1958, S. E. Flanders. Host: *Chrysomphalus* on citrus. DeBach code IA.7:1, S&R 1805-II-4"; 3. (bar code) "UCRC ENT 004289".

The holotype, deposited in USNM, is in a very good condition except that flagellum of one antenna and the tip of one forewing are missing.

Paratypes. 9 specimens on 5 slides (UCRC), as follows: 1 female remounted in Canada balsam, labeled: "S and R 1805 *Aphytis spiniferus* Limeira, Brazil, Flanders, 3/21/58 IA.7:6 [DeBach code, in pencil]. 2 females remounted on to separate slides in Canada balsam, labeled (new labels): "Brazil, Limeira, 14.iv.1958, S. E. Flanders. Host: *Chrysomphalus aonidum* (L.) on citrus. DeBach code IA.7:9 (or 10), S and R 1814-III". 1 female in Hoyer's, labeled: "New Genus Near *Aphytis* Ex *Chrysomphalus ficus?* on palm. Rio de Janeiro, Brasil, 4/5/58, S. E. Flanders Coll. 1795-5 S&R 1800-7 [crossed out] IA.7:8 [DeBach code, in pencil] *A. spiniferus*". 5 females under 3 coverslips (1, 2, and 2) on the same slide in Hoyer's, labeled: "Limeira, Brazil, iii.26 1958. HOST *Chrysomphalus ficus* Coll. S. E. Flanders NO. R. 1806-III-1 IA.7:7 [DeBach code, in pencil].

NON-TYPE MATERIAL. Also remounted were 53 non-type specimens of *A. costalimai* from Argentina and Brazil [UCRC].

Aphytis cylindratus Compere

(*chrysomphali* species group)

Aphytis cylindratus Compere 1955: 303; Rosen and DeBach 1979: 604-607.

TYPE MATERIAL. Before remounting, there were no type specimens of this species in UCRC; the single Canada balsam slide with the entire syntype series of *A. cylindratus* was received from USNM. According to Compere (1955), the syntype series of this species consisted of 2 females and 2 males but Rosen and DeBach (1979), probably by mistake, indicated that there were 2 females and 3 males, which they used, along with other material, to redescribe *A. cylindratus*. The syntype slide [USNM] is labeled (in black and red ink as well as in pencil) as: 1. "*Aphytis cylindratus* Comp. COTYPES photo"; 2."*Aphytis* n. sp. Clausen's No. 1441 Ex A. duplex on persimmon "External parasite 1-5 under scale" Osaka, Japan, Jan. 5, 1921 VIA.22:1 COTYPES *cylindratus* n. sp.". This slide has one female under the coverslip on the left, here designated as a lectotype to avoid possible confusion regarding the type specimens of this species; 2 male paralectotypes under the right (uppermost) coverslip, and one specimen, also a paralectotype, under the coverslip in the center (on the bottom of the slide), whose sex is impossible to determine because the distal part of the metasoma is badly damaged. According to Compere (1955), it was a female but in our opinion it looks more like a male. We added the appropriate labels to mark clearly the lectotype and the paralectotypes ("*Aphytis cylindratus* Compere LECTOTYPE ♀ (under left coverslip) Des. S. V. Triapitsyn and J.-W. Kim PARALECTOTYPES: 2 ♂♂ (upper right coverslip) and 1 of unknown sex (bottom center, a female according to Compere, 1955)"). The lectotype female is in poor condition, completely dissected and, like other original syntype specimens, stained, apparently with fuchsin. The two paralectotype males are also dissected into many parts, but the remaining paralectotype specimen of an unknown sex is almost entire, but with a few body parts missing and the wings shriveled. The single lectotype/paralectotype slide of this species is deposited in USNM on permanent loan from UCRC.

NON-TYPE MATERIAL. Also remounted were 21 non-type specimens of *A. cylindratus* from Japan and 5 specimens identified as *A. ?cylindratus* from Brazil [UCRC].

Aphytis dealbatus Compere

(related to *vittatus* species group)

Aphytis dealbatus Compere 1955: 286-287; Rosen and DeBach 1979: 290-291.

TYPE MATERIAL. A slide, containing the holotype female and one paratype female of *A. dealbatus*, both uncleared and mounted in Hoyer's, was received on loan from USNM. They were probably remounted into Hoyer's earlier from gum mar, the original mounting medium (Compere 1955). According to Compere (1955), the holotypes of his species were to be deposited in USNM; Rosen and DeBach (1979) erroneously indicated its depository as UCRC.

Holotype female. Original label: "*Aphytis dealbatus* Compere HOLOTYPE and PARATYPE Ex *Lepidosaphes ulmi* on black willow. Placerville, California Aug. 26, 1952 T. Fisher, Coll. 1B.71:1 [in pencil] The larger specimen the HOLOTYPE". After remounting, the holotype is labeled as follows: 1. "*Aphytis dealbatus* ♀ Compere HOLOTYPE Remounted from Hoyer's into Canada balsam 2000 By M. Planoutene (UCR)"; 2. "U.S.A., California, Placerville, 26.viii.1952, T. Fisher. Host: *Lepidosaphes ulmi* (L.) on black willow DeBach code IB.17:1"; 3. (bar code) "UCRC ENT 004314". The holotype, deposited in USNM, is in a very poor condition, uncleared, and broken into many parts.

Paratype. Also remounted was the above-mentioned paratype female of *A. dealbatus* [UCRC], with the same label data as holotype.

Aphytis debachi Azim

(*chrysomphali* species group)

Aphytis debachi Azim 1963: 287-290; Rosen and DeBach 1979: 601-604.

TYPE MATERIAL. Before remounting, there were no type specimens of this species in UCRC other than the two females on one slide, collected at the Kyushu University, Fukuoka, Japan, apparently from the syntype series but not marked as such (Rosen and DeBach 1979). Another slide, containing a headless syntype female, mounted in Hoyer's, was received from USNM, where it was apparently deposited earlier by mistake.

Syntype female. Original labels: 1. "*Aphytis debachi* Azim, 1963 Co-type [on a green label glued to the first label] Rec'd from Yasumatsu. Inconsistent with description #2"; 2."LOC. Fukuoka, Kyushu, Japan DATED Aug. 27 1962 HOST *Parlatoria* sp. DET. BY Azim 1963 ON VIB.1:1 [in pencil] COLL. A. AZIM"; 3. (on the back of the slide) "[Kyushu] Fukuoka 27.VIII.1962 A. Azim"; 4. (on the back of the slide) "*Aphytis debachi* Azim, 1963 Host: *Parlatoria* sp.". After remounting, the syntype is labeled as follows: 1. "*Aphytis debachi* ♀ Azim SYNTYPE Remounted from Hoyer's into Canada balsam 2000 By M. Planoutene (UCR)"; 2. "Japan, Kyushu, Fukuoka, 27.viii.1962, A. Azim. Host: *Parlatoria* sp. Note on orig. slide "Rec'd from Yasumatsu. Inconsistent with description" DeBach code VIB.1:1"; 3. (bar code) "UCRC ENT 004794". The syntype, deposited in UCRC, lacks the head and all the appendages except bases of the left forewing and the left hind wing.

Also remounted from Hoyer's were the two above-mentioned females, deposited in UCRC, which most probably also belong to the syntype series but are not marked as such, with similar original label data as holotype except for the collection date, which was 15.vii.1961. After remounting, they are labeled as follows: 1. "*Aphytis debachi* ♀ Azim Det. D. Rosen 1970 Remounted from Hoyer's into Canada balsam 2000 By M. Planoutene (UCR)"; 2. "Japan, Fukuoka, Kyushu Univ., 15.vii.1961, A. Azim. Host: *Parlatoria* sp. Comment - not labeled as cotype, but possibly belongs to syntype series of Azim. DeBach code VIB1:2"; 3. (bar codes) "UCRC ENT 004792" and "UCRC ENT 004793", respectively.

NON-TYPE MATERIAL. Also remounted were 3 non-type specimens from Hong Kong, China, determined as *A. ?debachi* [UCRC].

Aphytis desantisi DeBach and Rosen

(*proclia* species group)

Aphytis desantisi DeBach and Rosen 1976: 543; Rosen and DeBach 1979: 424-426.

TYPE MATERIAL. Before remounting, there were 11 paratype specimens in UCRC mounted on 10 slides in Hoyer's. In addition, the holotype female was received from USNM; it also was mounted on a slide in Hoyer's.

Holotype female. Original labels: 1. (in red and black ink and in pencil) "NAME *Aphytis desantisi*, n. sp. HOLOTYPE ♀ DET. DeB., DR 1969 COLL. Guyot NO IID.13.1 Dept. Biol. Cont. Univ. Calif."; 2. (in black ink) "LOC. El Siambon, Tucuman Pr., Argentina DATE VI-6 1969 HOST *Aonidiella aurantii* DET. 196 ON oranges". After remounting, the holotype is labeled as follows: 1. "*Aphytis desantisi* ♀ DeBach and Rosen HOLOTYPE Det. DeBach and Rosen 1969 Remounted from Hoyer's into Canada balsam 2000 By M. Planoutene (UCR)"; 2. "Argentina, Tucumán, El Siambon, 6.vi.1969, collector: original slide says Guyot, but Rosen and DeBach 1979 say A. Teran. Host: *Aonidiella aurantii* on orange. DeBach code IID.13:1"; 3. (bar code) "UCRC ENT 004593". The holotype, deposited in USNM (on permanent loan from UCRC), is in excellent condition.

Paratypes. 11 females on separate slides (after remounting), with the same data as holotype, deposited as follows: 1 female [IMLA], 1 female [MLPA], 1 female [TAMU], 1 female [USNM], remaining 7 females in UCRC.

Aphytis equatorialis Rosen and DeBach

(*lingnanensis* species group)

Aphytis equatorialis Rosen and DeBach 1979: 564-567.

TYPE MATERIAL. Originally this species was described from 24 female and 46 male syntypes mounted in Hoyer's on two slides. These two slides were received from USNM. The specimens were in poor condition as the mounting medium was dry and darkened. In the process of remounting, only 67 syntype specimens were found, as follows.

Lectotype female, here designated to avoid confusion about the identity of this species and status of its type specimens after remounting. Original labels: 1. "*Aphytis equatorialis* n. sp. Syntypes, DR and DeB 1976, Coll. A. Vilardebo, No. 2, IVA.24: 2"; 2. "Ivory Coast, Africa, Letter, II-16 1971, *Aspidiotus destructor* on Avocado (in alcohol)". The lectotype female was originally mounted with other 10 female and 21 male syntypes. After remounting, the lectotype female is labeled as follows: 1. "*Aphytis equatorialis* ♀ Rosen and DeBach, SYNTYPE LECTOTYPE designated by J.-W. Kim and S. V. Triapitsyn. Det. Rosen and DeBach 1976 Remounted from Hoyer's into Canada balsam 2000 By M. Planoutene (UCR)"; 2. "Ivory Coast, 16.ii.1971, A. Vilardebo. Host: *Temnaspidiotus destructor* on avocado. No. on orig. slide "2" DeBach code IVA.24:1"; 3. (bar code) "UCRC ENT 004662". The lectotype, deposited in USNM (on permanent loan from UCRC), is in fair condition but undercleared, with the head detached from the body.

Paralectotypes. The aforementioned 10 females and 21 males [1 female and 1 male in BMNH, 1 male in CNCI, 1 female and 1 male in TAMU, 1 female and 1 male in USNM, remainder in UCRC], remounted from the same original slide as the lectotype (same label data). Also 11 female and 24 male paralectotypes [UCRC] were remounted from the other syntype slide with the same label data as lectotype except for DeBach's code (IVA.24: 2).

COMMENTS. According to Rosen and DeBach (1979), the syntype series of *A. equatorialis* consisted of 24 females and 46 males; however, 2 females and 1 male were not found in the process of remounting (probably, these authors miscounted the true number of syntypes). All the type specimens including the lectotype have distorted antennae and bodies. Therefore, it is not easy to distinguish this species based solely on short and robust antennae. The more reliable distinguishing features of *A. equatorialis* are non-elongated crenulae and immaculate mesosterna. Additionally, most specimens have the scutellar sensilla close to the anterior scutellar setae. Many females also have a relatively short ovipositor. Males also have distinctly (anteriorly) positioned scutellar sensilla, as well as round and non-elongated crenulae. Less extensive stippling on the ventral side of the metasoma in males is another character that can be used for recognition of this species.

Aphytis fabresi DeBach and Rosen

(*vittatus* species group)

Aphytis fabresi DeBach and Rosen 1976: 542; Rosen and DeBach 1979: 259-262.

TYPE MATERIAL. Before remounting, there were 29 paratype specimens in UCRC mounted on 18 slides in dried Hoyer's. In addition, two slides were received from USNM, one containing the holotype female of *A. fabresi* and the other two males, the allotype and a paratype; they all also were mounted in dried Hoyer's.

Holotype female. Original labels: 1. (in red and black ink) "NAME *Aphytis fabresi*, n. sp. HOLOTYPE ♀ DET. D. Rosen 1972 COLL. G. Fabres NO. D Dept. Biol. Cont. Univ. Calif."; 2. (in black ink and pencil) "LOC. New Caledonia DATE HOST IA.17:1 DET. 19 ON caught in a sticky trap". After remounting, the holotype is labeled as follows: 1. "*Aphytis fabresi* ♀ DeBach and Rosen HOLOTYPE Det. D. Rosen 1972 Remounted from Hoyer's into Canada balsam 2000 By M. Planoutene (UCR)"; 2. "New Caledonia, Noumea, No date, collected by G. Fabres. No host information. No. on orig. slide "D". Note on orig. slide "caught in a sticky trap".

DeBach code IA.17:1"; 3. (bar code) "UCRC ENT 004791". The holotype, deposited in USNM (on permanent loan from UCRC), is in a very good condition, with flagellum of one antenna detached.

Paratypes. Allotype male [USNM], labeled as follows (after remounting): "New Caledonia, Noumea, 20.iii.1969, G. Fabres, Host: ? Note on orig. slide "bird-lime trap; all in alcohol" DeBach code IA.17:2". Other paratypes (1 female and 1 male in CNCI, 1 female and 1 male in TAMU, 1 female and 1 male in USNM, remaining in UCRC). DeBach and Rosen (1976) only specified the country for the type series of this species, also stating that their new species was described from numerous specimens. Later, Rosen and DeBach (1979) redescribed it from the entire type series, which included the holotype, the male allotype, as well as other paratype specimens (10 females and 22 males) for which more detailed label data were indicated; these data are not provided here to avoid unnecessary repetition. Out of 32 paratypes of *A. fabresi*, we remounted 8 female and 22 male specimens; the whereabouts of the remaining 2 female paratypes is unknown (both of these females were collected by G. Fabres in Noumea, New Caledonia, one on a bird-lime (sticky) trap 20.iii.1969 and the other, without the collection date, was remounted into Hoyer's by D. Rosen and P. DeBach from H. Compere's original Canada balsam slide).

Aphytis faurei Annecke

(*chilensis* species group)

Aphytis faurei Annecke 1963: 343-345; Rosen and DeBach 1979: 344-346.

TYPE MATERIAL. There was one paratype female on a slide (in Canada balsam) in UCRC; it was used by Rosen and DeBach (1979) to redescribe this species. Original labels: 1. "*Aphytis faurei* n. sp. ♀ - PARATYPE det. Annecke

IC.19:1 [DeBach's code]"; 2. ♀ N T669 South Africa: Ngwavuma, Natal, ix.1961, D. P. Annecke With *Africaspis chionaspiformis* and *Melanaspis corticosa* On leafless tree" [UCRC]. The holotype of *A. faurei* is deposited in SANC (Annecke 1963).

Aphytis fisheri DeBach

(*lingnanensis* species group)

Aphytis fisheri DeBach 1959: 362; Rosen and DeBach 1979: 561-564.

TYPE MATERIAL. DeBach (1959) wrote (p. 362): "Types of *Aphytis fisheri* are to be deposited in the U. S. National Museum, and paratypes in the British Museum and in the collection of the Department of Biological Control, University of California at Riverside. Two slides bearing 20 to 30 adults of both sexes, as well as some pupae mounted in Hoyer's medium, will be deposited in each institution". In fact, the two "type" slides deposited in USNM, listed as having 39 specimens (with "USNM No. 64851" [type number] written in a different hand than the original labels), are labeled as "cotypes" in P. DeBach's handwriting. Thus, P. DeBach's designation of "paratypes" in the original description was an invalid designation as the specimens on the two "paratype" slides, which remained in UCRC, and also on the additional two "paratype" slides deposited by him in BMNH, are in fact all syntypes. Apparently, there was confusion following the description of *A. fisheri* regarding which two of the six slides belonging to the type series were intended to be the "types" intended for USNM. Most likely P. DeBach himself did not mark them originally, forgot about his own earlier published designations, and therefore later chose to label all six slides as "cotypes", forgetting that two of them had been designated by him as "types" and the remaining four as "paratypes". Indeed, Rosen and DeBach (1979) mentioned only "syntypes" or "the type series" of this species. No other slides of *A. fisheri* in UCRC bear paratype labels. The two UCRC syntype slides were sent to USNM on permanent loan (Schauff 1985), but when we retrieved them for remounting, we didn't receive either of the two original USNM syntype slides, which are in poor condition (the specimens are mounted in badly deteriorated Hoyer's medium under 2 coverslips), or the two original syntype slides presently deposited in BMNH (J. S. Noyes, personal communication). Remounting of the two original UCRC syntype slides resulted in 29 female, 7 male, and 18 pupal syntypes, from which we designate a lectotype.

Lectotype female, here designated to clarify the status of the type specimens of this species. Original labels: 1. (in red ink) "*Aphytis fisheri* n. sp. Ex. *Aonidiella aurantii* on Rose Taunggyi and Heho, S. Shan States, Burma, Dec 21-25, 1956, Coll. P. DeBach"; 2. (in red ink) "Cotypes, reared in insectary on *Aspidiotus hederae*, Riverside Calif. Oct. 1958, Hoyer's medium, VA.20:2, DeBach". After remounting, the lectotype is labeled as follows: 1. "*Aphytis fisheri* ♀ DeBach SYNTYPE Remounted from Hoyer's into Canada balsam 2000 By M. Planoutene (UCR)"; 2. "USA, Ca., Riverside, UCR insectary culture on *Aspidiotus nerii*. Host: *Aonidiella aurantii* on rose "cotypes from culture" coll. by P. DeBach at Taunggyi and Heho, South Shan States, Burma, 21-25.xii.1956. DeBach code VA.20:2"; 3. (bar code) "UCRC ENT 004452". 4. "LECTOTYPE designated by J.-W. Kim and S. V. Triapitsyn". The lectotype, deposited in USNM (on permanent loan from UCRC), is in excellent condition.

Paralectotypes. From the aforementioned syntype slide, remounted were (besides the lectotype) 18 females as well as 5 males and 6 pupae [1 female in CNCI, 1 female in TAMU, 1 female and 1 male in USNM, 1 female and 1 male in ZIN, remainder in UCRC]. Additionally, 11 females as well as 2 males and 12 pupae [UCRC] were remounted from the second syntype slide with the same label data but bearing a different DeBach's code (VA.20:1). Also the 39 specimens on the two original syntype (or "type") slides in USNM and the uncounted specimens on the two original syntype (or "paratype") slides in BMNH.

COMMENTS. *Aphytis fisheri* can be separated from *A. melinus* DeBach by the characteristically unpigmented pupae. Thus, examining the pupal stage is required for the recognition of these two species because their adults are almost identical. The comparison table (p. 562 in Rosen and DeBach 1979) between *A. fisheri* and *A. melinus* is not reliable because the morphological characters they used are variable and not clearly differentiated. The antenna and mesoscutal setae of *A. fisheri* are darker than those of *A. melinus*. The antenna of *A. fisheri* is also relatively longer.

Aphytis funicularis Compere
(*funicularis* species group)

Aphytis funicularis Compere 1955: 279, 282-283; Quednau 1964: 279, 282-283; Rosen and DeBach 1979: 647-650.

TYPE MATERIAL. The type specimens of this species were originally all mounted in Canada balsam (Compere 1955) and the three paratype slides were in UCRC, among them one well-preserved slide with six male paratypes remounted from Canada balsam into Hoyer's (Rosen and DeBach 1979). Another Canada balsam slide with the holotype female, allotype male, and 6 paratypes (3 females and 3 males) was received on loan from USNM. According to Compere (1955), this species was described from the series of 30 specimens including the holotype and the allotype. Not all of them are found in UCRC and USNM.

Holotype female. Original labels: 1. "*Aphytis funicularis* Compere - Holotype 3rd from right. Allotype 7th from right. All others Paratypes. VIC.14:1 [DeBach's code, in pencil] *funiculus* [sic, in pencil]"; 2. "4 and 4 [in pencil, referring to the number of males and females on this slide] Ex. *Chionaspis chaetachmae* on "umkavoti" Durban, Natal, Aug. 28, 1925. E. W. Rust, Coll.". The holotype, deposited in USNM, is uncleared but otherwise in good condition; its position on the coverslip is indicated by a circle made in India ink.

Paratypes. Allotype male and also another 3 female and 3 male paratypes, under the same coverslip and on the same slide with the holotype [USNM]. Six males on slide (in Hoyer's): 1. "*Aphelinus* [crossed out] 6 ♂ *Aphytis abnormicornis* [a manuscript name, crossed out] Ex. *Chionaspis chaetachmae* on "umkavoti" Durban, So. Africa, Sept. 9, 1925. Rust. VIC.14:2 [DeBach's code, in pencil] Paratypes"; 2. "*Aphytis funicularis*" [UCRC]. One male on slide (in Canada balsam): 1. "*Aphelinus* [crossed out] ♂ *Aphytis abnormicornis* H. C. [a manuscript name, crossed out] Ex. *Diaspis rhusae* Camp's Bay, C. P. So. Africa, July 12, 1925. Rust. VIC.14:3 [DeBach's code, in pencil] Paratype"; 2. "*Aphytis funicularis* Comp." [UCRC]. One female on slide (in Canada balsam): 1. "Acc. No. 3 Rust's B3 *Aphytis abnormicornis* [manuscript name, crossed out] Host *Diaspis rhusae* July 17, 1925. Camp's Bay, C. P. *clavatus* [another manuscript name, partially covered by label #2] Coll. E. W. Rust. Det. Compere and Gahan"; 2. "*Aphytis funicularis*"; 3. "Drawing made Nov. 7, 31 1 ♀ VIC.14:4 [DeBach's code, in pencil]" [UCRC].

NON-TYPE MATERIAL. Also remounted from three slides were 16 female and 2 male non-type specimens of *A. funicularis* from Durban, Natal, South Africa, mentioned by Quednau (1964) and borrowed by J.-W. Kim from SANC.

Aphytis gordoni DeBach and Rosen
(*funicularis* species group)

Aphytis gordoni DeBach and Rosen 1976: 545; Rosen and DeBach 1979: 653-655.

TYPE MATERIAL. Before remounting, there were 28 paratype specimens in UCRC on 16 slides, all mounted in Hoyer's. In addition, the holotype female and allotype male, mounted in Hoyer's on the same slide with 3 additional paratypes of *A. gordoni*, were received from USNM.

Holotype female. Original labels: 1. (in red and black ink) "NAME *Aphytis* ♀ *gordoni* DeB. and Rosen HOLOTYPE ♀ upper r[ight] ALLOTYPE ♂ upper l[eft] DET. DR 1975 COLL. Cheng, S. K. No. R72-78 Div. Biol. Cont. Univ. Calif."; 2. (in black ink) "LOC. Hong Kong 72-89 DATE X-8 1972 HOST *Unaspis citri* V1C.18:1 [DeBach's code] ON Lemon ♂ #1-3 ♀ # 5, 6". After remounting, the holotype is labeled as follows: 1. "*Aphytis gordoni* ♀ DeBach and Rosen <u>HOLOTYPE</u> Remounted from Hoyer's into Canada balsam 1999 By M. Planoutene (UCR)"; 2. "U.S.A., CA, Riverside Co., Riverside U.C.R. Insectary. Orig. Hong Kong, 8.x.1972, S. K. Cheng. Host: *Unaspis citri* (Comstock) on lemon. DeBach code VIC.18:1, S and R 72-78"; 3. (bar code) "UCRC ENT 003708". The holotype, deposited in USNM (on permanent loan from UCRC), is in good condition and almost complete, just some parts of the middle leg are missing.

Paratypes. Allotype male [USNM] as well as 1 female and 2 male paratypes, with the same label data as holotype [USNM]. Also other paratypes [UCRC]: China, Hong Kong, ex. *Unaspis citri* (Comstock): originally collected ix-x.1971 by S. K. Cheng, 7 ♀, 6 ♂ (on citrus), then obtained by A. G. Selhime in Florida via the USDA Lab. in Moorestown, New Jersey, USA; 29.xi.1971, S. K. Cheng, 2 ♀ (on citrus); 21.vii.1972, S. K. Cheng, 3 ♀ (on tangerine); 7.viii.1972, S. K. Cheng, 1 ♀ (on tangerine); 24.ix.1972, S. K. Cheng, 2 ♀, 2 ♂ (on citrus); 1.x.1972, S. K. Cheng and C. T. Ho, 1 ♀, 1 ♂ (on citrus); 31.x.1972, S. K. Cheng, 1 ♀ (on citrus); 15.xi.1972, S. K. Cheng, 1 ♀, 1 ♂ (on citrus).

According to DeBach and Rosen (1976), the type series of *A. gordoni* consisted of 22 females and 14 males, including the holotype and the allotype. Of those, we remounted 33 specimens (20 females and 13 males). The paratype slides also included 6 pupae (also remounted), which were not part of the type series according to the original description.

Aphytis griseus Quednau

(related to *proclia* species group)

Aphytis griseus Quednau 1964: 100, 102-103; Rosen and DeBach 1979: 414-416.

TYPE MATERIAL. According to the original description, this species was described from 29 female and 30 male syntypes (Quednau 1964). Before remounting, there were 3 syntype slides in UCRC; an additional 5 slides were received on loan from SANC. All original slides had the Hoyer's mountant almost completely dark so that the specimens were scarcely visible even with good illumination. The lectotype female, designated by Rosen and DeBach (1979), was marked by a circle on one of the UCRC slides (Fig. 7).

Lectotype female. Original labels on the lectotype slide: 1. (red) "*Aphytis griseus* n. sp. Quedn. ♀ Syntypes Lectotype [circle pointed by an arrow]"; 2. "*Nelaspis exalbidaa* (Cock.) *Aloë arborescens* Pretoria 3.5.61 IID.1:3 [in pencil, DeBach's code]". After remounting, the lectotype female is labeled as follows: 1. "*Aphytis griseus* Quednau Lectotype female Remounted from Hoyer's into Canada balsam 2000 by M. Planoutene (UCR)"; 2. "South Africa, Pretoria, 3.v.1961, F. W. Quednau. Host: *Nelaspis exalbida* (Cockerell) on *Aloe arborescens* DeBach code IID.1:3"; 3."UCRC ENT 004788 [bar code]". The lectotype, deposited in USNM (on permanent loan from UCRC), is complete and in a fairly good condition.

Paralectotypes. 4 females [2 in UCRC, 2 in USNM], remounted from the lectotype slide, with the same label data; 6 males [3 in UCRC, 3 in USNM], remounted from the original syntype slide (DeBach code IID.1:4), same data; 7

males [6 in UCRC, 1 in TAMU], remounted from the original syntype slide (DeBach code IID.1:1), same data except the collection date is 7.iv.1961; 4 males [SANC], remounted from the original syntype slide, labeled: "1. (red) "*Aphytis griseus* n. sp. Quednau ♂ Syntypes F. W. Quednau det."; 2. "*Nelaspis exalbidaa* on *Aloë* Pretoria 20.ii.1962 Manting coll." (note that according to the original description, all the syntypes including those collected on 20.ii.1962 were collected by F. W. Quednau himself); 10 females [SANC], remounted from the original syntype slide labeled in the same way as the previous slide; 4 males [SANC], remounted from the original syntype slide, labeled: "1. (red) "*Aphytis griseus* Quedn n. sp. ♂ Syntypes"; 2. "*Nelaspis exalbidaa* (Cock.) *Aloë arborescens* Pretoria 7.iv.1961"; 9 females [SANC], remounted from the original syntype slide, labeled: "1. (red) "*Aphytis griseus* Quedn. ♀ Syntypes"; 2. "*Nelaspis exalbidaa* (Cock.) *Aloë arborescens* Pretoria 3.v.1961"; 4 females [SANC], same data as the previous slide, but this slide also had 3 pupae (2 females and 1 male) which were not mentioned as syntypes by Quednau (1964). Altogether, remounted were 28 out of 29 females and 21 out of 30 males belonging to the original syntype series. The whereabouts of the other paralectotypes (1 female and 9 males) is unknown to us.

Aphytis haywardi (De Santis)
(*vittatus* species group)

Marietta haywardi De Santis 1948: 146-149.
Aphytis haywardi (Blanchard): DeBach and Rosen 1976: 541; Rosen and DeBach 1979: 277-279.

TYPE MATERIAL. Before remounting, there was 1 female (apparently a paratype) specimen in UCRC on a slide (retained from the paratype series borrowed from Argentina). It was remounted in Hoyer's from the original Canada balsam slide (Rosen and DeBach 1979). We remounted it back into Canada balsam. Original labels: 1. "*Marietta haywardi* Blanch/De Santis [in pencil] ♀ Concordia, E. R. *Lecaniodiaspis* s. *Baccharis* 12.XI.19.34. K. Hayw."; "NAME *Aphytis* # 3 ♀ *haywardi* (Blanchard) - ?PARATYPE ♀ remounted IX-1973 Div. Biol. Cont. Univ. Calif.".

Aphytis hispanicus (Mercet)

(*proclia* species group)

Aphelinus maculicornis var. *hispanica* Mercet 1912b: 81-82.
Aphytis hispanicus (Mercet): Rosen and DeBach 1979: 390-394.

TYPE MATERIAL. The holotype of *A. hispanicus* was listed by Rosen and DeBach (1979) as being in MNMS; however, it was among the UCRC types of *Aphytis* spp. received by us from USNM for remounting and curation; probably this holotype specimen had been sent there by mistake. After completion of this study we returned this holotype to its proper depository in MNMS.

The holotype female of *Aphelinus maculicornis* var. *hispanica* Mercet is on a slide, mounted apparently in Hoyer's; because the condition of this specimen was very good, we refrained from remounting it (and we did not have a permission from the Museo de Ciencias Naturales in Madrid to do that anyway). The original labels are as follows: 1. "MUSEO DE MADRID LAB. DE ENTOMOL *Aphelinus maculicornis* v. *hispanica* Tipo [circled in pencil] Mercet"; 2. "MUSEO DE MADRID LAB. DE ENTOMOL S/. *Parlatoria* en Naranjo Valencia 27-XI-911 IIB.1:1 [DeBach's code, in pencil]"; 3. [red, on the bottom side of the slide]

"MUSEO DE MADRID LAB. DE ENTOMOL *Aphytis hispanicus* Mt. Holotipo ♀
Det. Ferrière".

NON-TYPE MATERIAL Also remounted from Hoyer's into Canada balsam
were 49 non-type specimens of *A. hispanicus* as well as 42 specimens identified by
D. Rosen and P. DeBach as "*A. ?hispanicus*" or as *A.* near *hispanicus*" [UCRC].

Aphytis holoxanthus DeBach
(*lingnanensis* species group)

Aphytis holoxanthus DeBach 1960: 704-705; Rosen and DeBach 1979: 548-552.

TYPE MATERIAL. Before the beginning of the project, there were numerous
syntype specimens in UCRC on 3 slides, all mounted in dry Hoyer's. The "lectotype"
female, mounted on the same slide with unmarked 45 females and 17 males, was
later received from USNM. This "lectotype" specimen was designated by Rosen and
DeBach (1979), and an inscription on that slide is in D. Rosen's hand. However, this
"lectotype" female was invalidly designated because this specimen was not part of
the syntype series designated in the original description by DeBach (1960), who
mentioned 9 syntype slides (3 each in BMNH, UCRC, and USNM). The "lectotype"
slide from USNM was also originally labeled differently from the 9 true syntype
slides. Therefore, after remounting, we designate a new lectotype of *A. holoxanthus*
from the true syntype specimens of this species from one of the 3 UCRC syntype
slides, as follows.

Lectotype female, here designated to clarify the status of the type specimens of
this species. Original labels: 1. "*Aphytis holoxanthus* DeBach, Syntypes, Reared on
Aspidiotus hederae in lab, VB.1:2"; 2. "Originally from Hong Kong – Cultured in
Israel – Then in Riverside, Calif., mount – July 10, 1960, Ex. *Chrysomphalus
aonidum* on citrus, See – Ann. Ent. Soc. Amer. 53 (6) 704, 1960". After remounting,

the lectotype is labeled as follows: 1. "*Aphytis holoxanthus* DeBach ♀ SYNTYPE Remounted from Hoyer's into Canada balsam 2000 By V. Berezovskiy (UCR)"; 2. "U.S.A., Ca, Riverside Co., UCR Insectary culture, 20.vii.1960. Orig. China, Hong Kong (orig. host: *Chrysomphalus aonidum* (L.), via Israel culture, Host: *Aspidiotus nerii* Bouché "see Ann. Ent. Soc. Am. 53(6): 704, 1960, DeBach code VB.1:2"; 3. (bar code) "UCRC ENT 004175"; 4. "LECTOTYPE des. by J.-W. Kim and S. V. Triapitsyn". The lectotype, deposited in USNM (on permanent loan from UCRC), is complete and in excellent condition.

Paralectotypes. 30 females, 6 males, 4 pupae, and 2 other specimens of an undetermined sex (the genitalia are missing), all on individual slides, as well as another slide with only a head, probably belonging to one of the above-mentioned specimens (there is one without a head), all remounted from the same slide as the lectotype and bearing the same label data [all in UCRC]. Additional paralectotypes (43 females, 17 males, 21 pupae, and 1 specimen of an undetermined sex [1 female and 1 male in BMNH, 1 female and 1 male in CNCI, 1 female and 1 male in TAMU, 2 females and 2 males in USNM, 1 female and 1 male in ZIN, remainder in UCRC]) were remounted from the other two UCRC syntype slides, with the same label data as lectotype but marked with different DeBach's codes (VB.1:4 and VB.1:5). There are also 6 additional original syntype slides in USNM and BMNH (3 in each collection), each slide containing numerous uncounted paralectotypes in badly deteriorated Hoyer's medium under 2 coverslips.

NON-TYPE MATERIAL. Also remounted were 117 specimens identified by D. Rosen and/or P. DeBach as *A. ?holoxanthus* or "*A.* nr. *holoxanthus*".

Aphytis hyalinipennis Rosen and DeBach
(related to *vittatus* species group)

Aphytis hyalinipennis Rosen and DeBach 1979: 285-287.

TYPE MATERIAL. The single Hoyer's slide with the entire type series of *A. hyalinipennis*, mounted together with a male *Metaphycus* sp. (Encyrtidae), was received from USNM.

Holotype female. Original labels: 1. (in black and red ink) "NAME *Aphytis hyalinipennis* n. sp. PARATYPE ♀ ALLOTYPE ♂ HOLOTYPE ♀ (lowest) DET. DR 1972 COLL. G. Fabres. Div. Biol. Cont. Univ. Calif."; 2. (in black ink) "LOC. Noumea New Caledonia DATE VIII. 1971 HOST Armored scale insect, undet. ON *Casuarina equisetifolia* IB.15:1 [DeBach's code]". After remounting, the slide with the holotype female is labeled as follows: 1. "*Aphytis hyalinipennis* ♀ Rosen and DeBach <u>HOLOTYPE</u> Det. D. Rosen 1973 Remounted from Hoyer's into Canada balsam 1999 By M. Planoutene (UCR)"; 2. "New Caledonia, Noumea, viii.1971, G. Fabres. Host: undetermined armored scale insect on *Casuarina equisetifolia* DeBach code IB.15:1"; 3. (bar code) "UCRC ENT 004310". The holotype, deposited in USNM (on permanent loan from UCRC), is in good condition and almost complete, just a foreleg is missing.

Paratypes. Also remounted were the allotype male [USNM] and 1 female paratype [UCRC], both with the same label data as holotype.

Aphytis ignotus Compere

(unassigned to species group)

Aphytis ignotus Compere 1955: 300-301; Rosen and DeBach 1979: 699-701.

TYPE MATERIAL. This species was described from 14 females and 7 males, including the female holotype and the male allotype, all collected at Botanical

Gardens, Sydney, Australia, on September 27, 1927 (Compere 1955). The single Canada balsam slide at UCRC contains 3 female and 2 male paratypes of this species marked so by H. Compere himself (in his handwriting). The original labels are as follows: 1. "*Aphytis ignota* [later corrected in blue ink to *ignotus*] Compere and Gahan [Gahan's name later crossed out in blue ink]. 3 ♀ and 2 ♂ paratypes Boiled in KOH VID.16:3 [in pencil, DeBach's code]"; 2. "Taken ovipositing in *Chrysomphalus rossi* on *Phoenix canariensis* in Botanical Gardens, Sydney, N.S.W. Sept. 27, 1927 H. Compere, Coll.". As with other H. Compere's original slides, the holotype and allotype had been mounted in Canada balsam and deposited by him in USNM (Compere 1955). However, they are missing from USNM and their present whereabouts are unknown.

NON-TYPE MATERIAL. Rosen and DeBach (1979) incorrectly mentioned a holotype female, an allotype male, and a paratype female of *A. ignotus*; however, these three specimens in UCRC were not part of the Compere's type series of this species. They labeled the Hoyer's slide containing the three aforementioned specimens, collected by S. E. Flanders in Sydney, Australia, on 21 April 1931 as follows: "L ♀ = HOLOTYPE *Aphytis ignota* 1 ♂ Allotype (L) 1 of ♀ Holotype, other paratype Number 2 Gahan letter". These were remounted in Canada balsam onto individual slides and relabeled accordingly, indicating their non-type status.

Aphytis immaculatus Compere

(*chrysomphali* species group)

Aphytis immaculatus Compere 1955: 307-308; Rosen and DeBach 1979: 610-613.

TYPE MATERIAL. This species was described from 11 females and 1 male, all designated by Compere (1955) as "cotypes", and mounted under three coverslips on the same slide. This slide was received from USNM, where it was deposited on permanent loan from UCRC (Schauff 1985) although that was unnecessary because Compere (1955) intended to deposit all primary types of his *Aphytis* species in USNM. These Hoyer's-mounted syntypes were not remounted because of their poor state: the 7 stained specimens (6 females and 1 male under two coverslips on the left side of the slide are all dissected and still are in the original Canada balsam), and the five female specimens on the right side of the slide were originally also mounted in Canada balsam but were unstained (Compere 1955); they were later remounted into Hoyer's (Rosen and DeBach 1979). The most complete specimen out of these 5 females is here designated as lectotype to avoid any confusion about the identity of the type specimens of this species.

Lectotype female [USNM]. The original labels are as follows: 1. "VIB.17:1 [in pencil, DeBach's code] COTYPES [highlighted in red] 5 ♀♀ unstained 6 ♀♀ stained *Aphytis taiwanensis* [later crossed-out in blue ink] n. sp. Compere = "Y" insectary notes *immaculatus* [later added in blue ink on the side of the label]"; 2. [on the bottom side of the slide] "*immaculatus* [later added in blue ink] *Aphytis taiwanensis* Comp. COTYPES [underlined in red] Ex. *Lepidosaphes* from Formosa on *Agalma*. Coll. T. C. Maa. Random sample taken by H. C. on Nov. 6, 1952 over. photos [in red ink]". The following two labels were added in the course of this study: 3. "LECTOTYPE ♀ PARALECTOTYPES 10 ♀♀, 1 ♂ *Aphytis immaculatus* Compere, 1955 Des. S. Triapitsyn 2004"; 4. "LECTOTYPE ♀ (circled)". The lectotype is in good condition, well cleared in Hoyer's, with one middle leg missing.

Paralectotypes. The above-mentioned 10 females and 1 male on the lectotype slide in USNM.

NON-TYPE MATERIAL. Remounted from Hoyer's into Canada balsam were 9 non-type specimens of *A. immaculatus* from China, including one specimen from the original material of this species (but not designated as a syntype), collected in Taiwan in November 1952 [UCRC].

Aphytis japonicus DeBach and Azim

(*chrysomphali* species group)

Aphytis japonicus DeBach and Azim 1962: 3-7; Rosen and DeBach 1979: 613-616.

TYPE MATERIAL. Before remounting, there were no marked type specimens of this species in UCRC, although present were several Hoyer's slides with numerous specimens of *A. japonicus* from the original collections of A. Azim in Fukuoka, Japan, during 1959-1961. Their status remains unclear as they were apparently a part of the original material of this species (DeBach and Azim 1962), but were not marked as syntypes by these authors. Three original syntype slides were received for curation from USNM. One of the slides contained a "holotype" female and the other an "allotype" male, both invalidly designated from the original syntype series by Rosen and DeBach (1979), who thus effectively designated a lectotype (i.e., the specimen marked by them as the "holotype"). The paralectotypes thus are all other "paratypes" examined by D. Rosen and P. DeBach including the "allotype" male, as well as all other syntypes deposited by the authors of this species in BMNH, USNM, and the collection of Entomological Laboratory of Kyushu University, Fukuoka, Japan.

Lectotype female. Original labels of the syntype slide containing the lectotype of *A. japonicus* are as follows: 1. (in blue ink) "Name *Aphytis japonicus* DeBach + Azim Syntypes ♀ ♀ + ♂ ♂ [in red ink] Det. ♀ HOLOTYPE [later added in black ink] 19 Mount. Hoyers By DeBach 1961 Dept. Bio. Contro. U. Calif."; 2. (in blue ink) "Loc. Fukuoka Japan VIB.19:3 [DeBach's code, in pencil] Date May 1961 Ex. *Chrysomphalus bifasciculatus Euonymus japonica* Collector A. Azim". After remounting, the slide with the lectotype female is labeled as follows: 1. "*Aphytis japonicus* ♀ DeBach and Azim <u>HOLOTYPE</u> - <u>No!</u> LECTOTYPE ♀ Effect. des. Rosen and DeBach 1979 Labeled by S. Triapitsyn 2004 Remounted from Hoyer's into Canada balsam 1999 By M. Planoutene (UCR)"; 2. "Japan, Fukuoka, v.1961, A. Azim. HOST: *Chrysomphalus bifasciculatus* Ferris On *Euonymus japonicus* DeBach code VIB.19:3 orig. a syntype"; 3. (bar code) "UCRC ENT 003919". The lectotype, deposited in USNM (on permanent loan from UCRC), is in bad condition, poorly cleared and missing all the wings except for one incomplete forewing, two pieces of which are detached from the body.

Paralectotypes. Also remounted were 12 female [8 in UCRC, 4 in USNM] and 2 male paralectotypes [UCRC, USNM] from the same original slide as the lectotype as well as 1 female and 1 male paralectotypes from the separate slide, with the same data as lectotype (the male was inappropriately marked by D. Rosen and P. DeBach as an "allotype"), as well as 2 female paralectotypes from another slide, also with the same data as lectotype.

Aphytis lepidosaphes Compere
(*chrysomphali* species group)

Aphytis lepidosaphes Compere 1955: 307; Rosen and DeBach 1979: 607-610.

TYPE MATERIAL. This species was described from numerous female and male specimens, all designated as syntypes, and mounted on 2 slides. These slides were received from USNM. After remounting of the slide, which contained 30 unstained specimens under one coverslip (according to Compere 1955), the most complete specimen of those is here designated as lectotype to avoid confusion about the identity of the type specimens of this species.

Lectotype female. The original labels are as follows: 1. "*Aphytis lepidosaphes* Compere"; 2. "photos [in red ink] *Aphytis* "X" IB1.1 [in pencil, unknown code] Ex.

Lepidosaphes beckii on citron melon C. E. S. Insectary Original stock from Formosa - coll. by Gressitt and his coolies or Maa or Djou - December 1950 VI.B.4:1 [DeBach's code, in pencil] COTYPES". After remounting, the slide with the lectotype female is labeled as follows: 1. "*Aphytis lepidosaphes* ♀ Compere COTYPE - *Aphytis lepidosaphes* Compere LECTOTYPE Des. by S. Triapitsyn 2004 Remounted from Hoyer's into Canada balsam 2000 By M. Planoutene UCR"; 2. "U.S.A., CA, Riverside Co., Riverside U.C.R. Insectary, orig. China, Formosa (Taiwan), xii.1950, J. L. Gressitt and Y. W. Djou? HOST: *Lepidosaphes beckii* (Newman) on citron melon. DeBach code VIB.4:1"; 3. (bar code) "UCRC ENT 003694". The lectotype, deposited in USNM (on permanent loan from UCRC), is in a rather poor condition but almost complete, with a part of one antenna detached.

Paralectotypes. 22 females, 3 males, and 1 specimen of an undetermined sex [1 female and 1 male in BMNH, 1 female and 1 male in CNCI, 1 female and 1 male in TAMU, 1 female and 1 male in USNM, 1 female and 1 male in ZIN, remainder in UCRC], remounted from the same slide as the lectotype (thus, the total number of specimens on that slide was 27, not 30 as was mistakenly indicated in the original description). Also apparently 5 females and 9 males [UCRC], dissected and stained, mounted in what looks like Canada balsam, on the other original syntype slide under two coverslips. The original labels of this slide are as follows: 1. "*Aphytis lepidosaphes* Compere"; 2. "*Aphytis lepidosaphius* [this specific manuscript name was later crossed out in pencil] Propagated on *Lepidosaphes beckii* Orange Co. Insectary by R. Smith Sept. 8, 1952 VI.B.4:2 [DeBach's code, in pencil] COTYPES". According to Compere (1955), that insectary was located in Anaheim, Orange Co., California.

NON-TYPE MATERIAL. Also remounted from Hoyer's into Canada balsam were 3 specimens identified as *A. ?lepidosaphes* from Hong Kong [UCRC].

Aphytis lingnanensis Compere

(*lingnanensis* species group)

Aphytis lingnanensis Compere 1955: 303-305; Rosen and DeBach 1979: 533-539.

TYPE MATERIAL. Prior to this project, there were about 30 specimens of *A. lingnanensis* in good quality Hoyer's mounts on 6 slides, all marked as "cotypes". It is obvious that those "cotype" markings were added onto the original labels later by someone else, but definitely not by H. Compere himself because they resemble D. Rosen's handwriting. In addition, we obtained 6 other slides from USNM from the same series, all of which were labeled as *Aphytis* "A" in H. Compere's handwriting. The species name "*lingnanensis*" was later added onto these slides. Neither Compere (1955) nor Rosen and DeBach (1979) were clear about the original type material of this species. Compere (1955) did not mention any type specimens at all. Rosen and DeBach (1979) mentioned the "type series" of *A. lingnanensis* in UCRC and also 3 slides from the syntype series that had been deposited in BMNH. According to A. Polaszek (personal communication), there are 5 slides in BMNH, each containing many specimens of *A. lingnanensis*; four of these slides are labeled as "cotypes". After thoroughly reviewing all the specimens of *A. lingnanensis* available to us, we conclude that only 16 slides (6 in UCRC, 6 in USNM, and 4 in BMNH) could be attributed to the original type series of Compere's *A. lingnanensis* beyond reasonable doubt. First, all these specimens were likely mounted and labeled by H. Compere himself (some of them were stained with fuchsin). Second, the label data matches Compere's remarks (p. 304 in Compere 1955) about the origin of UCR's insectary culture of this species: "1... the stock propagated in the insectary and subsequently established in California under the name "*Aphytis* A", traces back to parents imported into California from South China in November, 1947. The shipment was made by J.

Linsley Gressitt". Indeed, many of the slides were labeled as "China, Gressitt collection, Propagated on *A. hederae* on potato, Insectary C.E.S.". Third, many of these slides bear H. Compere's inscriptions (in pencil) for the drawings he made for the original description of *A. lingnanensis*.

None of the specimens of the syntype series has been remounted from Hoyer's into Canada balsam during the course of this project because most of them are dissected into numerous body parts and thus could be lost during the remounting process. Because we could not find the exact specimen from which the drawing of the body (with one pair of the wings attached, fig. 13, p. 304 in Compere 1955) was made (most probably it was a composite drawing based on several specimens), we select the most complete and well-positioned specimen on the second slide as lectotype.

Lectotype female, here designated to clarify the status of the type specimens of this species, with the following labels: 1. "*Aphytis* (A), from China, Gressit, Coll., Propagated in insectary C.E.S. on *Aspidiotus hederae* on potato, Oct. 18, 1948, H. Compere =*lingnanensis* [added later in blue ink] COTYPE [added later in red ink], IVA.1:2, ♀ # 1-2 [added later in pencil]"; 2. "*Aphytis lingnanensis* Compere PARALECTOTYPES des. by J.-W. Kim and S. V. Triapitsyn 2004"; 3. "*Aphytis lingnanensis* Compere ♀ LECTOTYPE des. by J.-W. Kim and S. V. Triapitsyn 2004". The lectotype, deposited in USNM (on permanent loan from UCRC), is circled in black ink. It is the fourth bottom specimen under the right coverslip (there are 4 females under the right coverslip and 3 females under the left coverslip). The lectotype female is in good condition, with a tarsus of one of the forelegs missing.

Paralectotypes. The total of 96 females and 18 males on 12 slides in UCRC and USNM, as follows: 1) 6 females under two coverslips (3+3) on the same slide with the lectotype [USNM]; 2) 12 females under one coverslip on slide labeled: "*Aphytis* (A) China, Gressit, Coll. Propagated in insectary C.E.S. on *Aspidiotus hederae*, Nov. 3, 1948, Drawing made Mar. 1952 IVA.1:1 COTYPE" [UCRC]; 3) 4 females and 1 male under two separate coverslips on slide labeled: "*Aphytis* (A), China, Gressitt, Coll. Propagated on *A. hederae* on potato. Insectary, C.E.S. Nov. 3, 1948 =*lingnanensis* drawing ♂ antenna COTYPE IVA.1:3" [UCRC]; 4) 5 females under one coverslip and 4 detached heads under another coverslip on slide labeled: "*Aphytis* (A), China, Gressitt Coll., on *A. hederae* on potato. Insectary C.E.S. Nov. 3, 1948 =*lingnanensis* COTYPE IVA.1:4" [USNM]; 5) 18 females (8+10) under two separate coverslips on slide labeled: "*Aphytis* (A), China, Gressitt, Coll Propagated on *Aspidiotus hederae* on potato. In Insectary, C.E.S. Nov. 3, 1948 =*lingnanensis* COTYPE IVA.1:5" [UCRC]; 6) 14 females (6+8) under two coverslips on slide labeled: "*Aphytis* (A) China. Gressit, Coll. Propagated on *A. hederae* on potato - Insectary C.E.S., Nov. 3, 1948, *A. lingnanensis* COTYPE IVA.1:6" [UCRC]; 7) 3 females and 7 males (2+2+3) under 4 separate coverslips on slide labeled: "*Aphytis* (A) =*lingnanensis* Propagated on *Aonidiella aurantii* on potato. Insectary C.E.S., Apr. 18, 1949, H. Compere 4 pairs setae propodeum Oct 6 ♂♂ IVA.1:7" [UCRC]; 8) 17 females and 12 males (5+12, 12 females under the same coverslip with the males) under two coverslips on slide labeled: "*Aphytis* (A) =*lingnanensis* Ex. *Aspidiotus hederae* on potato - C.E.S. Insectary. These being colonized at rate of 1500 per day Bartlett, Coll. Apr. 12, 1949, IVA.1:8 [USNM]"; 9) 4 females under the same coverslip on slide labeled: "*lingnanensis Aphytis* (A) Propagated on *A. hederae*, C.E.S. Insectary, Apr. 14, 1949 photo made IVA.1:9" [UCRC]; 10) 4 females under one coverslip on slide labeled: "*Aphytis* (A) *lingnanensis* Propagated on *A. hederae* C.E.S. Insectary Apr. 14, 1949 originally from Gressitt - China IVA.1:10 [USNM]"; 11) 4 females under one coverslip on slide labeled: "*Aphytis* (A) *lingnanensis* Propagated on *Aspidiotus hederae* C.E.S. Insectary Apr. 14, 1949 originally from Gressitt - China IVA.1:11" [UCRC]; 12) 5 females under three separate coverslips (2+1+2) on slide labeled: "*Aphytis* (A) *lingnanensis* Propagated on *Aonidiella aurantii* on potato Insectary, C.E.S. Apr. 18, 1949 Mounted by H. Compere

IVA.1:12" [UCRC]. The uncounted specimens on the four additional original syntype slides of *A. lingnanensis* in BMNH are also designated paralectotypes.

NON-TYPE MATERIAL. There are two males on a separate slide in UCRC that represent the original material received from China and propagated in UCR quarantine (collected on August 10, 1949), labeled by H. Compere as *"Aphytis* A?". Because of that question mark we do not regard them as part of the original syntype series of *A. lingnanensis*. Also we found two other slides, each with two coverslips, labeled as follows: *"Aphytis* "A" type pupae ♂ Propagated on red scale by Ted Fisher in Insectary C.E.S. Riverside. Mar. 17, 1950 Parent stock from Hughes Ranch, Texas. Coll. by DeBach Dec. 1949". However, no pupae can be found mounted on these slides (rather, 2 females are on the first slide and numerous males on the other, labeled as "stippled"); these specimens apparently were not part of the type series because Compere (1955, p. 305) mentioned some of the stocks of *Aphytis* from Texas as only provisionally assigned to *A. lingnanensis*, also noting that the males had "...pointlike flecks in the integument on the venter of the abdomen, producing a stippled effect in stained specimens".

Remounted from Hoyer's into Canada balsam were 270 non-type specimens identified as *A. lingnanensis* and also 616 specimens identified as *A. ?lingnanensis* or *A.* near *lingnanensis*.

Aphytis longicaudus Rosen and DeBach
(unassigned to species group)

Aphytis longicaudus Rosen and DeBach 1979: 681-684.

TYPE MATERIAL. Before remounting, there were 3 paratype specimens of this species in UCRC, mounted in Hoyer's on 2 slides. The slide that contained the holotype female, the allotype male, and the paratype female, all also mounted in Hoyer's, was received from USNM.

Holotype female. Original labels: 1. (in red ink) "NAME *Aphytis longicaudus* n. sp. HOLOTYPE ♀ (up-r) ALLOTYPE ♂, PARATYPE ♀ DET. D.R. 1975 COLL. S. E. Flanders Div. Biol. Cont. Univ. Calif."; 2. (in black ink) "LOC. Taipo Hong Kong DATE XII-11 1953 HOST *Pseudaonidia trilobitiformis* DET 19 ON citrus VID.4:1 [DeBach code, in pencil]". After remounting, the holotype is labeled as follows: 1. "*Aphytis longicaudus* ♀ Rosen and DeBach <u>HOLOTYPE</u> Det. D. Rosen 1975 Remounted from Hoyer's into Canada balsam 2000 By M. Planoutene (UCR)"; 2. "Hong Kong, Taipo, 11.xii.1953, S. E. Flanders. Host: *Pseudaonidia trilobitiformis* (Green) on citrus. DeBach code VID.4:1"; 3. (bar code) "UCRC ENT 004324". The holotype, deposited in USNM (on permanent loan from UCRC), is in fair condition but missing a clava of one antenna and one forewing; the head and the remaining forewing are detached from the body.

Paratypes. Also remounted from the same slide with the holotype were the allotype male [USNM] and the paratype female [UCRC], as well as two pupae [UCRC], mentioned by Rosen and DeBach (1979). The other three paratypes at UCRC were not remounted from Hoyer's because of the danger of their disintegration; the original labels are as follows: 1. "*Aphelinus* Silv. # 14 [in pencil] ♀ ♂ on *Pseudaonidia duplex* C. Kowloon"; 2. [in pencil] "sp.?? 2 ex. *Pseudaonidia duplex* see slide 1"; 3. "NAME *Aphytis longicaudus* n. sp. PARATYPE ♀ ♂ DET. D.R. 1975 COLL. F. Silvestri NO. VID.4:2 [DeBach's code] Div. Biol. Cont. Univ. Calif.", 1 female and 1 male on the same slide. Also a female paratype on a separate slide with similar label data except for F. Silvestri's number (#13) and DeBach's code (VID.4:3).

Aphytis luteus (Ratzeburg)

(unassigned to species group)

Coccobius luteus Ratzeburg 1852: 196.

Aphytis luteus (Ratzeburg): Rosen and DeBach 1979: 486-489.

TYPE MATERIAL. Before remounting, there were 2 type (originally syntype) specimens of this species in UCRC, mounted in Hoyer's on 2 slides. The first slide contained the lectotype female, designated by Bouček (1964), and erroneously labeled by D. Rosen and P. DeBach as "holotype". The second slide contained the paralectotype female, erroneously labeled by D. Rosen and P. DeBach as "?paratype". Both were received by them from the Ratzeburg collection in DEI, and remounted from the original card mounts (Rosen and DeBach 1979). Unfortunately, the original labels from the card points were probably destroyed in the course of that first remounting; these were in Ratzeburg's handwriting: "*luteus* Rtz" and "*Coccus Pini*" (Rosen and DeBach 1979). After we remounted these two specimens from Hoyer's into Canada balsam, both the lectotype and the paralectotype were intended for return to their original depository (DEI); however, attempts to contact the curator(s) there failed and thus they at least temporarily remain in UCRC.

Lectotype female. "Original" (i.e., in this case D. Rosen and P. DeBach's) labels: 1. "NAME *Aphytis* (TYPE) *luteus* HOLOTYPE [in red ink] (*Coccobius luteus* Ratzb.) DET. Ratzb. 1848 COLL. Ratzeburg Remounted from point into Hoyer's VII-15-70 Dept. Biol. Cont. Univ. Calif."; 2. "LOC. Eberswalde Germany DATE 1848 HOST "*Coccus pini*" IIIC.4:1 [DeBach code, in pencil] DET 19 ON = "Z". After remounting, the lectotype is labeled as follows: 1. "*Coccobius luteus* ♀ Ratzeburg LECTOTYPE Designated by Bouček (1964) [*Aphytis luteus* (Ratzeburg)] Remounted from Hoyer's into Canada balsam 2000 By M. Planoutene (UCR)"; 2. "Germany, Eberswalde, 1848. Host: *Coccus pini* Note on orig. slide "holotype; = "Z" Remark - wrongly labeled as holotype DeBach code IIIC.4:1"; 3. (bar code) "UCRC

ENT 004775". The lectotype is in fair condition but missing parts of one antenna and several leg segments; the head and both forewings and one hind wing are detached from the body.

Paralectotypes. The above-mentioned female, with the same label data as lectotype. Also another female in DEI, erroneously listed by Taeger et al. (2005) as a "holotype".

NON-TYPE MATERIAL. Also remounted from Hoyer's into Canada balsam were 63 non-type specimens of *A. luteus* from Corsica (France) and Greece [UCRC].

Aphytis maculatipennis (Dozier)
(*vittatus* species group)

Marietta maculatipennis Dozier 1933: 88-89.
Aphytis maculatipennis (Dozier): Rosen and DeBach 1979: 241-244.

TYPE MATERIAL. This species was described from numerous specimens; the holotype female and the allotype male were deposited in USNM (Dozier 1933). The species was redescribed by Rosen and DeBach (1979) from 5 female and 8 male paratypes, remounted by them into Hoyer's from one original slide onto 8 slides; the original Canada balsam slide was apparently received by them from USNM and retained in UCRC. We remounted 5 of these 8 Hoyer's slides; the remaining 3 slides contain dissected parts of the remaining 3 paratype specimens.

Paratypes. Labels on the original (H. L. Dozier's) slide: 1." *Marietta maculatipennis* Dozier ♂ & ♀ Reared from *Diaspidiotus* sp. nov. on mahogany. Cote Plage, Haiti June 28-1931 H. L. Dozier"; 2. (red) "*Marietta maculatipennis* Dozier IA.5:5 [DeBach's code, in pencil, added later- S.T.] ♀ ♂ forewings other specimens remounted D. Rosen 1968 ParaType No. U.S.N.M. [crossed out later, apparently by

D. Rosen]". After remounting on to individual slides in Canada balsam, the 5 female and 7 male paratypes [UCRC, USNM] are labeled as follows: 1. "*Marietta maculatipennis* [female or male symbol] Dozier PARATYPE [*Aphytis maculatipennis*] Det. (Dozier) D. Rosen 1968 Remounted from Hoyer's into Canada balsam 2000 By M. Planoutene (UCR)"; 2. "Haiti, Côte Plage, 28.vi.1931, H. L. Dozier. Host: *Diaspidiotus* sp. nov. on mahogany DeBach code IA.5:1"; 3. (a bar code).

Aphytis maculatus (Shafee)
(related to *vittatus* species group)

Syediella maculata Shafee 1970: 144.

Aphytis malayensis Rosen and DeBach 1979: 294-296.

Aphytis maculatus (Shafee): Hayat 1998: 83-85.

TYPE MATERIAL. *Aphytis malayensis*, a synonym of *A. maculatus* (Hayat 1998), was described from a single holotype specimen (Rosen and DeBach 1979) on slide in Hoyer's, which was received from USNM.

Holotype female. Original labels: 1. "NAME *Aphytis malayensis*, n. sp. HOLOTYPE ♀ [highlighted in red] n. sp. nr. *vittatus* [in pencil] DET. DR 1970 COLL. Chua Tock Hing NO N-16 TYPE [in red ink] Dept. Biol. Cont. Univ. Calif."; 2. "LOC. Malaya DATE 19 HOST DET 19 ON I IB.22:1 [DeBach code, in pencil]". After remounting, the holotype is labeled as follows: 1. "*Aphytis malayensis* ♀ Rosen and DeBach HOLOTYPE Det. D. Rosen 1970 Remounted from Hoyer's into Canada balsam 2000 By M. Planoutene (UCR)"; 2. "Malaysia. No date information. From a sample of *Saissetia nigra* (Nietner). Collector: Chua Tock Hing. No. on orig. slide "N-16", Note on orig. slide "N. sp. near *vittatus*" DeBach code IB22:1"; 3. (bar code) "UCRC ENT 004361". The holotype, deposited in USNM (on permanent loan from UCRC), is in good condition but missing parts of one antenna and one middle leg; the head is detached from the body.

Aphytis mandalayensis Rosen and DeBach
(related to *vittatus* species group)

Aphytis mandalayensis Rosen and DeBach 1979: 296-299.

TYPE MATERIAL. This species was described from two female specimens, the holotype and a paratype (Rosen and DeBach 1979). The holotype, mounted on slide in Hoyer's, was received from USNM. The paratype specimen in UCRC, which was dissected (Rosen and DeBach 1979), has not been remounted.

Holotype female. Original labels: 1. [In red ink] "NAME *Aphytis mandalayensis* n. sp. HOLOTYPE ♀ DET. DR 1975 COLL. P. DeBach NO. S and R 1688 Div. Biol. Cont. Univ. Calif."; 2. [In blue ink] "S and R 1688 *Aphytis* n. sp. Mandalay, Burmah 2/7/57 DeBach, Coll. Ex. red on rose? IB.24:1 [DeBach code, in pencil] HOLOTYPE [in pencil]". After remounting, the holotype is labeled as follows: 1. "*Aphytis mandalayensis* ♀ Rosen and DeBach HOLOTYPE Det. D. Rosen 1975 Remounted from Hoyer's into Canada balsam 2000 By M. Planoutene (UCR)"; 2. "Burma, Mandalay, 7.ii.1957, P. DeBach. Host: *?Aonidiella aurantii* (Maskell) on rose. DeBach code IB24:1 S&R 1688"; 3. (bar code) "UCRC ENT 004358". The holotype, deposited in USNM (on permanent loan from UCRC), is in a very good condition but incomplete (one antenna and flagellum of the other antenna are missing); the head is detached from the body.

Paratypes. The paratype female on slide [UCRC], labeled: 1. "NAME *Aphytis mandalayensis* n. sp. PARATYPE ♀ DET. DR 1975 COLL. P. DeBach NO. S and R 1688 Div. Biol. Cont. Univ. Calif."; 2. "*Aphytis* n. sp. S and R 1688 Red scale? on Rose Mandalay, Burma Feb. 7, 1957 DeBach, Coll. IB.24:2 [DeBach code, in pencil] PARATYPE [in pencil]".

Aphytis margaretae DeBach and Rosen

(*lingnanensis* species group)

Aphytis margaretae DeBach and Rosen 1976: 544; Rosen and DeBach 1979: 545-548.

TYPE MATERIAL. Before remounting, there were numerous paratype specimens in UCRC on 18 slides, all mounted in dry Hoyer's. We also received from USNM a slide containing the holotype female, the allotype male, and a paratype pupa of *A. margaretae*.

Holotype female. Original labels: 1. (in red ink) "*Aphytis margaretae*, n. sp. HOLOTYPE ♀, ALLOTYPE ♂, pupa, Det. DeBach, Coll. P. DeBach"; 2. (in black ink) "La Paz, Baja California, 23 Jun. 1966, *Diaspis echinocacti*, Det. DeB. 1966, on *Opuntia* sp. IVB.22:1". After remounting, the holotype is labeled as follows: 1. "*Aphytis margaretae* female DeBach and Rosen HOLOTYPE ♀ Det. DeBach 1966 Remounted from Hoyer's into Canada balsam 2000 By M. Planoutene (UCR) "2. "Mexico, Baja California Sur, La Paz, 23.vi.1966, P. DeBach. Host: *Diaspis echinocacti* (Bouché) on *Opuntia* sp. DeBach code IVB.22:1"; 3. (bar code) "UCRC ENT 004594". The holotype, deposited in USNM (on permanent loan from UCRC), is in excellent condition. There is a discrepancy in the collection date of the holotype between the original description of DeBach and Rosen (1976), who indicated it as

July 1966, and the later redescription of *A. margaretae* by Rosen and DeBach (1979), where they indicated the correct collection date (June 23, 1966).

Paratypes. The allotype male [USNM] and a paratype pupa [USNM] were remounted from the original holotype slide, with the same label data. Other paratypes include 265 females, 93 males, and 6 pupae as well as 12 specimens of an undetermined sex (due to bad quality of the specimens) [1 female and 1 male in BMNH, 1 female and 1 male in CNCI, 1 female and 1 male in TAMU, 5 females and 5 males in USNM, 1 female and 1 male in ZIN, remainder in UCRC]. All these paratypes were remounted from the 18 original slides, the labels of which match the data indicated by Rosen and DeBach (1979).

Aphytis mazalae DeBach and Rosen

(*chrysomphali* species group)

Aphytis mazalae DeBach and Rosen 1976: 544; Rosen and DeBach 1979: 616-620.

TYPE MATERIAL. Before remounting, there was a single slide of this species in UCRC, which contained two female paratypes mounted in Hoyer's. Another original slide, containing the female holotype and the male allotype (and also a pupa which was not part of the type series), was received from USNM.

Holotype female. Original labels: 1. [in red, blue, and black ink and in pencil] "♀ = 2, ♂ = 3 NAME *Aphytis mazalae* HOLOTYPE ♀, ALLOTYPE ♂ nr. *citrinus* DET. DeB 3/15 1966 COLL. P. Lin-Moo Peng NO. letter 2/15/66 Dept. Biol. Cont. Univ. Calif. ok"; 2. [In blue ink] "VIB.22:1 [DeBach's code, in pencil] Nr. Taipei, Taiwan DATE 14 FEB 1966 HOST *Chrysom-phalus ??aonidum* DET DeBach 1966 on citrus Host is ??". After remounting, the holotype female is labeled as follows: 1. "*Aphytis mazalae* ♀ DeBach and Rosen Holotype Det. DeBach

15.iii.1966 Remounted from Hoyer's into Canada balsam 2000 By M. Planoutene (UCR)"; 2. "Taiwan, near Taipei, 14.ii.1966, Lin-Moo Peng. Host: *Chrysomphalus aonidum* on citrus. Note on orig. slide "near *citrinus*, letter 2/15/66" DeBach code VIB22:1 S&R"; 3. (bar code) "UCRC ENT 004927". The holotype, deposited in USNM (on permanent loan from UCRC), is complete and in excellent condition.

Paratypes. The allotype male on slide [USNM], remounted from the same original slide as the holotype; also 2 females remounted from a separate original slide [UCRC], with the same label data as holotype.

NON-TYPE MATERIAL. Also remounted from Hoyer's into Canada balsam were 3 specimens determined by D. Rosen and P. DeBach as *A. mazalae* and 9 specimens determined by them as *A. ?mazalae* (all from Pakistan) [UCRC].

Aphytis melanostictus Compere

(related to *vittatus* species group)

Aphytis melanostictus Compere 1955: 287; Rosen and DeBach 1979: 291-294.

TYPE MATERIAL. This species was described from 90 specimens of both sexes (mostly males), from which the holotype female was selected (Compere 1955). The holotype was supposedly placed by the author in USNM (Compere 1955) but the USNM type catalog did not record that (T. Nuhn, personal communication). It is missing from USNM and its present whereabouts are unknown. It is quite likely, however, that this "holotype" has never been selected because apparently H. Compere himself forgot to label the type series of this species properly and later did not check what he wrote in the original description back in 1955, thus improperly labeling the specimens of the type series in UCRC as "cotypes". Indeed, five slides with the specimens belonging to the type series of *A. melanostictus* remained in UCRC, all of which were mounted in Canada balsam (uncleared) and are marked in

Compere's handwriting as "cotypes" (it is even quite possible that one of them actually is the unmarked holotype but that is impossible to determine for sure). The type locality indicated in the original description is Mira Loma, in Riverside County, California, USA (Compere 1955); the one indicated on the slides is Wineville, an archaic name of the location that is now within the boundaries of Mira Loma. The collection date indicated in the original description is March 15, 1936 (which is the date on three of the slides in UCRC) whereas the specimens on the two other slides in UCRC were collected on March 10, 1936 (thus they are indeed the paratypes).

Paratypes (as stated above, one of them may be the holotype) [all in UCRC]. 7 females, 12 males on slide labeled: 1. "*Aphytis melanostictus* Compere ♀ ♂ COTYPES [underlined in red] IB.19:4 [DeBach's code, in pencil] Ex. *Aspidiotus juglansregiae* on walnuts Wineville, Riverside Co., Calif. Mar. 15, 1936 S. E. Flanders, Coll."; 8 females, 22 males and one specimen of an unknown sex on one slide and 1 male on another slide, with the same label data (except for DeBach's code); also 5 males on one slide and 1 female, 1 male on another slide, with the same label data except the collection date (March 10, 1936), the latter are not marked as "cotypes" (the specific name "*melanostictus*" was added by H. Compere later in red ink), but retaining the original label as "*Marietta* n. sp.", which is hidden beneath the aforementioned newer labels on the four other slides.

Aphytis melinus DeBach
(*lingnanensis* species group)

Aphytis melinus DeBach 1959: 361-362; Rosen and DeBach 1979: 552-557.

TYPE MATERIAL. Before remounting, there was no marked type material of this species in UCRC. The three slides of *A. melinus*, which originally belonged to UCRC and were marked in P. DeBach's handwriting as "cotypes", were then received from USNM. All the specimens on these slides were mounted in dry Hoyer's. According to DeBach (1959), the type series of his *A. melinus* consisted of six slides, each bearing 20-30 specimens of both sexes and some pupae; USNM, UCRC, and BMNH were supposed to receive two slides each. Oddly, only the two slides deposited in USNM were considered by the author of this species as "types", whereas he treated the other specimens of the type series as "paratypes". However, both slides originally deposited in USNM were also labeled by P. DeBach as "cotypes", in exactly the same manner as the three (not two, apparently the third slide was originally intended for BMNH but instead was retained at UCRC) slides in UCRC. Therefore, all the specimens on the original six slides belonging to the type series of *A. melinus* should be considered syntypes. We selected a lectotype of *A. melinus* from the remounted syntype specimens from one of the original UCRC syntype slides, as follows.

Lectotype female, here designated to clarify the confused status of the type specimens of this species. Original labels (Fig. 11): 1. (in red ink) "*Aphytis melinus*, n. sp. Ex. *Aonidiella aurantii* + *citrina* on Rose + Citrus, New Delhi + Gurgaon, India, Lahore + Rawalpindi, Pakistan, Sept. 1956; April 1957, Coll. Angalet & DeBach"; 2. (in red ink) "Cotypes, Reared in insectary on *Aspidiotus hederae* Riverside, Calif. Oct. 1958, Hoyer's medium, from mixed culture DeBach, VA.1:3 [DeBach's code in pencil]". After remounting, the lectotype is labeled as follows (Fig. 12): 1. "*Aphytis melinus* ♀ DeBach SYNTYPE *Aphytis melinus* DeBach LECTOTYPE Des. by J.-W. Kim & S. Triapitsyn 2004 Remounted from Hoyer's into Canada balsam 2000 By M. Planoutene (UCR)"; 2. "California, Riverside, UCR insectary culture on *Aspidiotus nerii*, x.1958. Host: *Aonidiella aurantii* and *A. citrina* "cotypes from mixed culture" coll. in N. Delhi & Gurgaon, India and Lahore & Rawalpindi, Pakistan by Angalet in 1956, DeBach in 1957. DeBach code VA.1:3"; 3. (bar code) "UCRC ENT 004404". The lectotype, deposited in USNM (on permanent loan from UCRC), is in good condition and complete, with a forewing detached from the body.

Paralectotypes. 17 females, 6 males, and 4 pupae [1 female in BMNH, 1 female in CNCI, 1 female in TAMU, 2 females and 2 males in USNM, 1 female and 1 male in ZIN, remainder in UCRC], remounted from the same original slide as lectotype, with the same label data. Additionally, 36 females, 8 males, and 10 pupae [1 female and 1 male in AMUZ, 8 females and 3 males in USNM, 1 female in ZIN, remainder in UCRC] were remounted from the other two original UCRC syntype (or "paratype") slides with the same label data but bearing different DeBach's codes (IIA.1: 1 and IIA.1: 2, respectively). There are also 2 original syntype (or "type") slides in USNM, with "USNM No. 64852" [type number] written in a different hand than the labels, each containing numerous paralectotype specimens in badly deteriorated Hoyer's medium under 2 coverslips; these were not sent to us and thus have not been remounted. The uncounted specimens on the sixth original syntype slide, presently deposited in BMNH (J. S. Noyes, personal communication), are also designated paralectotypes.

NON-TYPE MATERIAL. Also remounted were 303 non-type specimens of *A. melinus* (including the voucher specimens of the original material from India and Pakistan, before the cultures were mixed together) and, additionally, 57 specimens identified as *A. ?melinus* or "*A.* near *melinus*".

COMMENTS. Coloration of the mature pupae of *A. melinus* (with black mesosomal sterna) is a good distinguishing feature from the unpigmented pupae of *A. fisheri*. Also, *A. melinus* has a paler antennal clava and usually somewhat paler setae on the mesosoma than *A. fisheri* or *A. holoxanthus*. However, some paralectotypes of *A. melinus* have a mixture of pale and dark setae on the mesosoma.

Aphytis merceti Compere
(*chilensis* species group)

Aphytis merceti Compere 1955: 299; Rosen and DeBach 1979: 341-344.

TYPE MATERIAL. This species was described by Compere (1955) from a series of 12 females and 4 males, all collected by E. W. Rust in South Africa. A larger part of that series, all mounted on separate slides in Canada balsam, including the holotype female, remained in UCRC, while two additional slides of the type series were received from USNM. According to Compere (1955), all the holotypes and allotypes of his species described in that paper were deposited in USNM; thus Rosen and DeBach (1979) erroneously indicated UCRC as depository of the holotype of *A. merceti* and also made a mistake (following H. Compere's mistake in labeling of the holotype and the paratypes) in identifying the true holotype specimen. In the original description, Compere (1955) made several mistakes in interpreting the label data of the type series of his *A. merceti*, indicating the incorrect collection dates and also the wrong number of specimens from one of the type localities (Newlands). Nevertheless, he unambiguously indicated that the holotype female and the allotype male were chosen from the two females and two males collected in Newlands, Cape Town (although there was another male specimen from the same location, labeled by him as a paratype). Indeed, there is a slide among these specimens on which H. Compere wrote "Holotype" in pencil, and it is the true holotype. Later, however, H. Compere himself erroneously labeled one of the slides with a paratype female of *A. merceti* from Rosebank, Cape Town, as holotype.

Holotype female. Original labels: 1. "*Aphelinus* ♀ HOLOTYPE Ex *C. rossi* Newlands, C. P. 1C.14:7 [DeBach's code, in pencil] II-14-23. Rust."; 2. "*Aphytis merceti* Compere and Gahan ["Gahan" is later crossed out in pencil] 1 ♀ paratype". A third label was added in the course of this project to clearly indicate the primary type status of this specimen: "*Aphytis merceti* Compere ♀ HOLOTYPE Det. S. Triapitsyn 2004". The holotype, deposited in USNM, is complete and in fair condition (although it is uncleared).

Paratypes. Allotype male [USNM], originally labeled: 1. "*Aphelinus* ♂ Y. C.14:2 [DeBach's code, in pencil] Ex *C. rossi* Newlands, C. P. Feb. 10-23. Rust."; 2. "*Aphytis merceti* Compere and Gahan [Gahan is later crossed out in pencil] ♂ 1 paratype [later crossed out in pencil] ALLOTYPE [in pencil]"; 3. [yellow] "allotype Rs'91". Other paratypes: 1 female [USNM], same data as above but collected 15.ii.1923; 1 male [UCRC], same data but collected 14.ii.1923; 1 male [UCRC], same data but collected 10.ii.1923 (incorrectly labeled by H. Compere as an allotype); 1 male [UCRC], same data but collected 14.ii.1923; 3 females [UCRC] on 3 slides, reared by E. W. Rust in Rosebank, Cape Province, South Africa, from a sample of *Gascardia destructor* (Newstead) (as "*Ceroplastes destructor*" on the original slides), as follows: 1 female, 18.ix.1924 ("boiled KOH"); 1 female, 19-24.ix.1924 ("Drawing made of this specimen"); 1 female, 20-24.ix.1924 (erroneously labeled by H. Compere as "*Aphytis merceti* Compere and Gahan ["Gahan" was crossed out later in pencil] 1 female Holotype"). Also 1 female and 3 males [UCRC] on 3 slides (2 males on the same slide), reared by E. W. Rust in Groote Schuur, Cape Province, South Africa, from a sample of *Melanaspis corticosa* (Brain) (as "*Chrysomphalus corticosus*" on the original slides), as follows: 1 female, 21.i.1925; 1 male, 31.i.1925; 2 males, 2.ii.1925. In the paratype list, Compere (1955) erroneously indicated 3 females and 1 male from the latter location.

NON-TYPE MATERIAL. Remounted from Hoyer's into Canada balsam were 3 non-type specimens of *A. merceti* from South Africa [UCRC].

Aphytis mimosae DeBach and Rosen
(unassigned to species group)

Aphytis mimosae DeBach and Rosen 1976: 545; Rosen and DeBach 1979: 694-696.

TYPE MATERIAL. Before remounting, there were two slides of this species in UCRC with two paratypes mounted in Hoyer's. The rest of the type series (the female holotype, the male allotype as well as two female and one male paratypes) were erroneously listed in UCRC by Rosen and DeBach (1979) but were later appropriately returned to SANC.

Paratypes [UCRC]. 1 female, originally labeled (in red and black ink and in pencil): 1. "♀ # 1 NAME *APHYTIS* n. sp. nr. *wollumbillae mimosae*, n. sp. PARATYPE ♀, DET. DR 1970 COLL. G. J. Snowball NO T-2395 Dept. Biol. Cont. Univ. Calif."; 2. "LOC. Pienaarspoort, Tvl. South Africa DATE 4 1966 HOST *Gascardia mimosae* DET 19 VID.12:3 [DeBach's code] ON *Acacia karroo*". After remounting, the paratype female is labeled as follows: 1. "*Aphytis mimosae* ♀ DeBach and Rosen Paratype Det. D. Rosen 1970 Remounted from Hoyer's into Canada balsam 2000 By M. Planoutene (UCR)"; 2. "South Africa, Transvaal, Pienaarspoort, iv.1966, G. J. Snowball. Host: *Gascardia mimosae* on *Acacia karoo* No. orig. slide "T-2395" Note on orig. slide "n. sp. nr. *wollumbillae*". Note "presumably erroneous host record, *Gascardia*" DeBach code VID.12:3 S&R"; 3. (bar code) "UCRC ENT 004763". Also 1 male, with the same original label data except for the number [T-2395 (a)].

Aphytis moldavicus Jasnosh
(*mytilaspidis* species group)

Aphytis moldavicus Jasnosh in Nikol'skaya and Jasnosh 1966: 207; Rosen and DeBach 1979: 475-476.

TYPE MATERIAL. Before remounting, there were 8 female paratypes of this species in UCRC, individually mounted on slides in Hoyer's. According to Rosen and DeBach (1979), these were mounted from alcohol-preserved samples, and the original labels (in Russian) were not preserved. The labels on the Hoyer's slides were mostly correct and agree with the original description, which is in Russian, except for the collector's name that was misspelled on two of these slides as "V. Tolizky". After remounting, the following, more appropriate, data labels were provided.

Paratypes [UCRC]. 4 females, labeled: "USSR, Moldavia [now Transdnyestr Republic (Moldova)], Dubossary, 2.vii.1958, V. Talitsky, coll. Host: *Lepidosaphes* sp. on *Syringia vulgaris* Note on orig. slide "vial 10 slides [a number follows], DeBach code IIID.1: [a number follows]". 2 females, labeled: "USSR, Moldavia [now Moldova], Kishinev, 7.vi.1960, V. Talitsky, coll. Host: *Lepidosaphes ulmi* (L.) Note on orig. slide "vial 9 slides [a number follows], DeBach code IIID.1: [a number follows]". 2 females, labeled: "USSR, Moldavia (now Moldova), Kishinev, 1961, V. Talitsky, coll. Host: *Lepidosaphes ulmi* (L.) DeBach code IIID.1: [a number follows]".

NON-TYPE MATERIAL. Also remounted from Hoyer's into Canada balsam were 3 non-type specimens of *A. moldavicus* from Tajikistan [UCRC].

Aphytis mytilaspidis (Le Baron)
(*mytilaspidis* species group)

Aphelinus mytilaspidis Le Baron 1870: 360-362.
Aphytis mytilaspidis (Le Baron): Rosen and DeBach 1979: 464-473.

TYPE MATERIAL. This species was redescribed by Rosen and DeBach (1979) from the "neotype series" of 39 females and 2 males, which included the neotype female. Before remounting, there was a slide labeled as "Neotypes" in UCRC, which contained the entire "neotype series", all mounted in Hoyer's. According to Rosen and DeBach (1979, p. 466), this slide was numbered "II A12:6" but that number was missing, apparently erased and changed to the standard DeBach's code (IIIA.1:1). The original "II A12" numbers, however, can be found on some other slides at UCRC with the same collection data; specimens on these slides, however, are not part of the "neotype series". Also present were 4 slides with 7 females and 1 pupa from the same collection event, labeled as "Neotype series" in D. Rosen's handwriting but not mentioned by Rosen and DeBach (1979) as part of that series.

Neotype female. Original labels: 1. [in red ink] "NAME *Aphytis mytilaspidis* (Le Baron) NEOTYPES DET. DR and DeB 1970 COLL. NO. R-65-11 Div. Biol. Cont. Univ. Calif."; 2. [In black ink] "LOC. Urbana Illinois DATED Feb 16 1965 HOST *Lepidosaphes ulmi* IIIA.1:1 [DeBach's code] DET BY ON Privet COLL. C. E. White"; 3. [in red ink] "Large ♀ 4TH ROW FROM RIGHT, 7 FROM TOP = NEOTYPE DET. DR 1973". After remounting, the neotype female is labeled as follows: 1. "*Aphytis mytilaspidis* ♀ (Le Baron) NEOTYPE Det. Rosen and DeBach 1970 Note: D. Rosen selected this specimen in 1973 Remounted from Hoyer's into Canada balsam 2000 By M. Planoutene (UCR)"; 2. "U.S.A., Illinois, Urbana, 16.ii.1965, C. E. White. Host: *Lepidosaphes ulmi* (L.) on privet. DeBach code IIIA.21:1, S&R R-65-11"; 3. (bar code) "UCRC ENT 004697". The neotype, deposited in USNM (on permanent loan from UCRC), is in good condition, complete, with both forewings detached.

NON-TYPE MATERIAL. Remounted from the same slide with the neotype were 38 females and 2 males [UCRC] that were part of the "neotype series" but erroneously labeled on the original slide as "neotypes"; however, none of these specimens has any type status.

Also remounted from Hoyer's into Canada balsam were 32 non-type specimens of *A. mytilaspidis* as well as 116 specimens identified by D. Rosen and/or P. DeBach as "*A. ?mytilaspidis*" or "*A.* nr. *mytilaspidis*" [UCRC].

Aphytis neuter Jasnosh and Myartseva
(related to *mytilaspidis* species group)

Aphytis neuter Jasnosh and Myartseva 1971: 36-37; Rosen and DeBach 1979: 490-492.

TYPE MATERIAL. Before remounting, there was 1 female paratype of this species in UCRC, mounted in a dry Hoyer's medium on a slide.

Paratype. 1 female [UCRC], originally labeled: "1 ♀ *Aphytis slavonicus* [this is a manuscript name] Jasnosh et Myartseva ex *Diaspidiotus slavonicus* Green USSR, Middle Asia, Dusanbe, June, 2, 1950, E. Borovkov. det. V. Jasnosh. Paratype. IIIC.7:1 [DeBach code, in pencil]". After remounting, this paratype female is labeled as follows: 1. "*Aphytis slavonicus* ♀ Jasnosh and Myartseva Det. V. Jasnosh ms name *A. neuter* J. and M. PARATYPE Remounted from Hoyer's into Canada balsam 2001 By V. Berezovskiy (UCR)"; 2. "USSR, Tajikistan, Dushanbe, 2.vi.1950, E. A. Borovkov. Host: *Quadraspidiotus slavonicus* (Green) on poplar. DeBach code IIIC.7:1"; 3. (bar code) "UCRC ENT 007011".

Aphytis nigripes (Compere)
(*vittatus* species group)

Marietta nigripes Compere 1936: 312.
Aphytis nigripes (Compere): Rosen and DeBach 1979: 268-270.

TYPE MATERIAL. The holotype female and a paratype female of this species, mounted on the same slide, are in USNM. The remaining two female paratypes are in UCRC, mounted individually on slides in Canada balsam, as follows.

Paratypes [UCRC]. 1 female, labeled: 1. "*Aphytis nigripes* (Compere) [in D. Rosen's handwriting] PARATYPE ♀ Coll. H. Compere IB.1:1 [DeBach's code, in pencil] Uncleared, balsam"; 2. "*Marietta nigripes* Comp. Cape Banks, Botany Bay. Sydney, N.S.W. Sept. 29, 1927 Coll. H. Compere. *nigripes* [in pencil] Paratype". 1 female, labeled: 1. "*Aphytis nigripes* (Compere) [in D. Rosen's handwriting] PARATYPE ♀ Coll. H. Compere IB.1:2 [DeBach's code, in pencil] Cleared, balsam"; 2. "*Marietta nigripes* Comp. PARATYPE Ex. coccid No. 101627 Cape Banks, Botany Bay - Sydney, N.S.W. Oct. 16, 1927 H. Compere (KOH)".

Aphytis obscurus DeBach and Rosen
(*vittatus* species group)

Aphytis obscurus DeBach and Rosen 1976: 541; Rosen and DeBach 1979: 279-282.

TYPE MATERIAL. Before remounting, there were 14 slides of this species in UCRC that contained 18 paratype specimens, all mounted in dry Hoyer's. Two other original slides, containing the female holotype and the male allotype, were received from USNM.

Holotype female. Original labels: 1. [in red and black ink] "NAME *Aphytis obscurus*, n. sp. HOLOTYPE ♀ DET. DR 1973 COLL. H. Zimmerman NO. Dept. Biol. Cont. Univ. Calif."; 2. [In black ink] "LOC. Martin Garcia Island, Argentina DATE VIII 1971 HOST *Diplacaspis echinocacti* DET 19 ON IB.11:1 [DeBach's code, in pencil]". After remounting, the holotype female is labeled as follows: 1. "*Aphytis obscurus* ♀ DeBach and Rosen Holotype Det. D. Rosen 1973 Remounted from Hoyer's into Canada balsam 2000 By M. Planoutene (UCR)"; 2. "Argentina, Martin Garcia Island, iii.1971, H. Zimmermann. Host: *Diaspis echinocacti* (Bouché). Rosen and DeBach 1979 noted "received from Dr. D. P. Annecke, Pretoria, South Africa; his code No. T3893". DeBach code IB.11:1"; 3. (bar code) "UCRC ENT 004360". The holotype, deposited in USNM (on permanent loan from UCRC), is in excellent condition but missing one foreleg; one antenna and flagellum of the other are detached from the head.

Paratypes. The allotype male is on a slide [USNM], remounted from a separate original slide, with the same label data as holotype; also other 15 female and 3 male paratypes [1 female in SANC, 1 female in MLPA, 1 female and 1 male in USNM, remainder in UCRC], remounted from 14 original slides in UCRC, with the same label data as holotype except the date on 7 of them is 1.viii.1971 (collector's last name was misspelled on some of the original slides); two of the females are incomplete (containing body parts only).

Aphytis paramaculicornis DeBach and Rosen

(*proclia* species group)

Aphytis paramaculicornis DeBach and Rosen 1976: 542-543; Rosen and DeBach 1979: 387-394.

TYPE MATERIAL. Before remounting, there were 3 slides of this species in UCRC that contained numerous paratype specimens (the total number of the paratypes was never counted by the authors of this species), all mounted in dry Hoyer's. Another original slide, containing the female holotype and the male allotype of *A. paramaculicornis*, was received from USNM. As original labeling of the type series of this species was inaccurate, we cite the more detailed information on its origin (Rosen and DeBach 1979, p. 387): "...reared in the insectary at Riverside on the cactus scale, *Diaspis echinocacti* (Bouché), originally from the olive scale, *Parlatoria oleae* (Colvée) collected in Iran in 1951 by A. M. Boyce) = "Persian *Aphytis*"); laboratory stock received from Albany, California, from material recovered from olive scale in the field in the San Joaquin Valley". This would be difficult to fit on a label, obviously, so we decided to copy the original labels on most of the paratypes, to avoid any additional, unnecessary confusion. The slides containing the holotype and the allotype were relabeled using the most important parts of the correct information, because the original data label was too inaccurate to be copied without making at least some necessary changes.

Holotype female. Original labels: 1. [in red and black ink] "NAME *Aphytis paramaculicornis* HOLOTYPE ♀ ALLOTYPE ♂ DET. DR 1972 COLL. lab reared NO. Rec'd from Berkely Div. Biol. Cont. Univ. Calif."; 2. [In black and red ink] "LOC. Persia (orig.) DATE VIII 1972 HOST ? ORIG. *P. oleae* [IIIB9.1 - crossed out] IIA.24:1 [DeBach's code, in pencil] DET 19 ON olive Lab stock received from Albany"; 3. [pink] "Holotype"; 4. "ALLOTYPE". After remounting, the holotype female is labeled as follows: 1. "*Aphytis paramaculicornis* ♀ DeBach and Rosen HOLOTYPE Det. D. Rosen 1972 Remounted from Hoyer's into Canada balsam 1999 By M. Planoutene (UCR)"; 2. "U.S.A., Ca., Riverside Co., Riverside, UCR Insectary, orig. from Iran, original HOST: *Parlatoria oleae* (Colvée) on olive, lab colony est. from stock received from Albany, California viii.1972. DeBach code IIA.24:1"; 3. (bar code) "UCRC ENT 003935". The holotype, deposited in USNM (on permanent loan from UCRC), is complete and in good condition.

Paratypes. The allotype male on slide [USNM], remounted from the same original slide as holotype; also other paratypes: 17 females and 16 males [1 female, 1 male in USNM; 1 female, 1 male in ZIN; remainder in UCRC], remounted from the 3 original slides in UCRC, with the following original labels: 1. "NAME *Aphytis paramaculicornis* PARATYPES DET. DR 1972 COLL. NO. Rec'd from Berkely Div. Biol. Cont. Univ. Calif."; 2. "LOC. Persia DATE VIII 1972 HOST ? DET 19 ON ? [subsequent DeBach's codes follow)"; 3. "Paratype". After remounting, these paratype females are individually labeled as follows: 1. "*Aphytis paramaculicornis* [a female or male symbol follows] DeBach and Rosen <u>PARATYPE</u> Det. D. Rosen 1972 Remounted from Hoyer's into Canada balsam 2000 By M. Planoutene (UCR)"; 2. "Iran (orig.) viii.1972 - UCR lab colony note on orig. slide "rec'd from Berkely [sic]; host ? DeBach code IIA.24: [2-4]".

Aphytis perplexus Rosen and DeBach

(*vittatus* species group)

Aphytis perplexus Rosen and DeBach 1979: 249-251.

TYPE MATERIAL. This species was described from 6 specimens: the female holotype, the 3 female paratypes, the male allotype, and a headless male paratype (Rosen and DeBach 1979). The single slide with the entire type series was received from USNM.

Holotype female. Original label [In red and blue ink]: "NAME *Aphytis perplexus*, n. sp. HOLOTYPE, ALLOTYPE and PARATYPES DET. BY DR 1972 1A.12:1 [DeBach code, in pencil] LOT NO. BRAZ 10 Holotype ♀ 3rd from top DR Dept Biol. Conn. Univ. Calif". The other (locality) label apparently peeled off the slide and was lost long ago. After remounting, the holotype is labeled as follows (locality and other data were taken from the original description): 1. "*Aphytis perplexus* ♀ Rosen and DeBach <u>HOLOTYPE</u> Det. D. Rosen 1972 Remounted from Hoyer's into Canada balsam 2000 By M. Planoutene (UCR)"; 2. "Brazil, Rio de Janeiro State, Rural University, 25.iv.1962. Host: *Hemiberlesia lataniae* (Signoret) on ornamental palm. DeBach code IA.12:1 S&R"; 3. (bar code) "UCRC ENT 003698". The holotype, deposited in USNM (on permanent loan from UCRC), is in very good condition but incomplete (one hind wing is missing); a middle leg and both hind legs are detached from the body.

Paratypes. The allotype male [USNM] as well as 3 female [2 in UCRC, 1 in USNM] and 1 male [UCRC] paratypes on separate slides, with the same label data as holotype.

Aphytis philippinensis DeBach and Rosen
(*proclia* species group)

Aphytis philippinensis DeBach and Rosen 1976: 543; Rosen and DeBach 1979: 397-400.

TYPE MATERIAL. Before remounting, there were 14 slides of this species in UCRC (all Hoyer's mounts) that contained about 45 paratype specimens. The total number of the paratypes was not counted by the authors of this species at the time of the original description but later was indicated by Rosen and DeBach (1979) as 42 females and 7 males. Another original slide, containing the female holotype and the male allotype of *A. philippinensis*, was received from USNM.

Holotype female. Original labels: 1. [in red, blue, and black ink] "NAME *Aphytis philippinensis* n. sp. HOLOTYPE ♀ ALLOTYPE ♂ DET. BY DR 1972 LOT NO. R-65-56 Original material DEPT BIOL. CONN. UNIV CALIF HOLOTYPE ♀ - ALLOTYPE ♂"; 2. [In blue ink] "LOC. Mandaue, Cebu, Philippines DATE June 14 1965 HOST *Chrysomphalus aonidum* DET 19 ON coconut palm Coll. V. G. Ortega IIB.21:1 [DeBach's code, in pencil]". After remounting, the holotype female is labeled as follows: 1. "*Aphytis philippinensis* ♀ DeBach and Rosen HOLOTYPE

Remounted from Hoyer's into Canada balsam 1999 By M. Planoutene (UCR)"; 2. "Philippines, Cebu, Mandaue, 14.vi.1965, V. G. Ortega. HOST: *Chrysomphalus aonidum* (L.) on coconut palm. S&R-65-56 DeBach code IIB.21:1"; 3. (bar code) "UCRC ENT 003918". The holotype, deposited in USNM (on permanent loan from UCRC), is in good condition but missing a hind wing and a tibia and tarsus of one of the hind legs; head, prosternum, forelegs, and one forewing are detached from the rest of the body.

Paratypes. The allotype male is on a slide [USNM], remounted from the same original slide as the holotype; also 37 female, 7 male paratypes [1 female in TAMU; 1 female and 1 male in USNM; 1 female in ZIN; remainder in UCRC], remounted from the above-mentioned 14 original slides, with the same label data as holotype except some new labels also copy the notes on the original slides, such as "Original material" or "nr. *hispanicus*; Filial".

NON-TYPE MATERTIAL. Also remounted from some of these paratype slides were 8 pupae, which were not designated as paratypes by DeBach and Rosen (1976). Remounted from Hoyer's into Canada balsam were 17 specimens from India identified by D. Rosen as "*Aphytis* nr. *philippinensis*".

Aphytis phoenicis DeBach and Rosen
(*mytilaspidis* species group)

Aphytis phoenicis DeBach and Rosen 1976: 543-544; Rosen and DeBach 1979: 485-486.

TYPE MATERIAL. Before remounting, there were 4 slides of this species in UCRC (all Hoyer's mounts) that contained about 100 paratype specimens. The total number of the paratypes was not counted by the authors of this species at the time of the original description but later (Rosen and DeBach 1979) was indicated as 106 females. Another original slide, containing the female holotype of *A. phoenicis*, was received from USNM.

Holotype female. Original labels: 1. [in red and black ink] "NAME *Aphytis phoenicis* HOLOTYPE ♀ DET. DR 1973 COLL. H. E. Martin NO. 5 Dept. Biol. Cont. Univ. Calif."; 2. "LOC. Aunaiza, Saudi Arabia DATE 7 III 1967 HOST *Parlatoria blanchardi* DET 19 ON date palm see Israeli sp. [in pencil] IIIC.1:1 [DeBach's code, in pencil]". After remounting, the holotype female is labeled as follows: 1. "*Aphytis phoenicis* ♀ DeBach and Rosen HOLOTYPE Det. D. Rosen 1973 Remounted from Hoyer's into Canada balsam 2000 By M. Planoutene (UCR)"; 2. "Saudi Arabia, Aunaiza, 7.iii.1967, H. E. Martin. HOST: *Parlatoria blanchardi* (Targioni-Tozzetti) on *Phoenix dactylifera*. No. on orig. slide "5", Note on orig. slide "see Israeli sp.". DeBach code IIIC.1:1"; 3. (bar code) "UCRC ENT 004589". The holotype, deposited in USNM (on permanent loan from UCRC), is complete and in good condition.

Paratypes (all collected at Riyadh-Erka, Saudi Arabia, by H. E. Martin from the same host as holotype). 5 females [UCRC] (30.iv.1967, remounted from 2 slides with DeBach's codes IIIC.1:2 and IIIC.1:3), 24 females [UCRC] (24.v.1967, remounted from a slide with DeBach's code IIIC.1:4), 74 females [1 female in TAMU; 7 females in USNM; 2 females in ZIN; remainder in UCRC] (11.vi.1967, remounted from a slide with DeBach's code IIIC.1:5).

NON-TYPE MATERIAL. Also remounted from Hoyer's into Canada balsam were 8 specimens of *A. phoenicis* from Israel.

Aphytis pilosus DeBach and Rosen

(unassigned to species group)

Aphytis pilosus DeBach and Rosen 1976: 545; Rosen and DeBach 1979: 684-687.

TYPE MATERIAL. Before remounting, there were 3 slides of this species in UCRC (all Hoyer's mounts) that contained 5 paratype specimens. According to the hand-written notes in the slide cabinet at UCRC, eight other slides, containing the remainder of the type series of this species (the holotype female, allotype male, 11 female and 3 male paratypes) were sent back to SANC because Rosen and DeBach (1979) mistakenly indicated UCRC as their sole depository.

Paratypes. After remounting, the 4 female and 1 male paratypes [UCRC] are labeled as follows: 1. "*Aphytis pilosus* DeBach and Rosen <u>PARATYPE</u> Det. D. Rosen 1969 Remounted from Hoyer's into Canada balsam 2000 By M. Planoutene (UCR)"; 2. "South Africa, Transvaal, Pienaarspoort, ii.1966, G. J. Snowball. HOST: *Gascardia mimosae* on *Acacia karroo*. No. on orig. slide "T2395 (B)". Note on orig. slide "n. sp. nr. *ciliatus*". DeBach codes VID.4:3, 4, and 6"; 3. (bar codes) "UCRC ENT 005403-005407".

NON-TYPE MATERIAL. Also remounted from Hoyer's into Canada balsam were 3 non-type specimens of *A. pilosus* from South Africa.

Aphytis pinnaspidis Rosen and DeBach

(*proclia* species group)

Aphytis pinnaspidis Rosen and DeBach 1979: 409-411.

TYPE MATERIAL. Before remounting, there were about 55 paratype specimens in UCRC on 9 slides, all mounted in Hoyer's. In addition, the holotype female and the allotype male, mounted on the same slide in Hoyer's, were received from USNM. According to Rosen and DeBach (1979), this species was described from the female holotype, the male allotype, as well as 40 female and 28 male paratypes from two different locations in Brazil.

Holotype female. Original labels: 1. (in blue and red ink) "NAME *Aphytis pinnaspidis* ♀ IIC.22:1 [DeBach code, in pencil] <u>TYPES</u> DET BY DR 1969 LOT NO. 31 DEPT BIOL. CONN. UNIV CALIF"; 2. (in blue ink) "LOC. Rural University, Rio de J. State, Brazil DATED June 11 1962 HOST *Pinnaspis strachani* Cooley DET BY 19 ON *Hibiscus* sp. COLL. DeBach". After remounting, the holotype is labeled as follows: 1. "*Aphytis pinnaspidis* ♀ Rosen and DeBach <u>HOLOTYPE</u> Remounted from Hoyer's into Canada balsam 1999 By M. Planoutene (UCR)"; 2. "Brazil, Rio de Janeiro State, Rural University, 11.vi.1962, P. DeBach. Host: *Pinnaspis strachani* Cooley on *Hibiscus* sp. DeBach code IIC.22:1"; 3. (bar code) "UCRC ENT 003915". The holotype, deposited in USNM (on permanent loan from UCRC), is in good condition, with both forewings detached from the body; one hind wing is missing.

Paratypes. The male allotype [USNM] was remounted from the same slide as the holotype (with the same label data). Also 16 more or less complete females and 13 more or less complete males as well as 2 slides with unassociated body parts, remounted from the original paratype slides, all with the same label data as holotype [UCRC]. Other paratypes [1 female in EMEC; 1 female in CNCI; 1 female in TAMU; 1 female, 1 male in USNM; remainder in UCRC]: 21 more or less complete females and 11 more or less complete males, also 8 slides with unassociated body parts, remounted from the original paratype slides, all labeled as follows: 1. "*Aphytis pinnaspidis* Rosen and DeBach <u>PARATYPE</u> Det. D. Rosen 1969 Remounted from Hoyer's into Canada balsam 2000 By M. Planoutene (UCR)"; 2. "Brazil, Vitoria, Pernambuco, 11.iv.1962 P. DeBach. Host: *Pinnaspis strachani* (Cooley) on *Solanum juribeba*".

NON-TYPE MATERIAL. Also remounted from some of the paratype slides were 5 non-type pupae, with the same label data as the two above-mentioned paratype series (3 and 2, respectively) [all in UCRC].

Aphytis riyadhi DeBach

(*lingnanensis* species group)

Aphytis riyadhi DeBach 1979: 131-138.

TYPE MATERIAL. Before remounting, there were the holotype and more than 100 paratype specimens in UCRC on 17 slides, all mounted in dry Hoyer's. According to DeBach (1979), the type series of *A. riyadhi* consisted of "numerous females".

Holotype female. Original labels: 1. "*Aphytis riyadhi* DeBach, HOLOTYPE, paratypes, holotype specimen is top right, specimen under right coverslip"; 2. "Diriyah (nr Riyadh), Saudi Arabia, V.26.1977. Host: *Aonidiella orientalis* on citrus". After remounting, the holotype is labeled as follows: 1. "*Aphytis riyadhi* DeBach ♀ HOLOTYPE, Remounted from Hoyer's into Canada balsam 2000 By M. Planoutene (UCR)"; 2. "Saudi Arabia, Dir'iyah (near Riyadh), 26.v.1977. Host: *Aonidiella orientalis* on citrus, C. E. Kennett, coll. Reared in UCR Quarantine (DeBach, 1979)"; 3. (bar code) "UCRC ENT 004270". The holotype, deposited in USNM (on permanent loan from UCRC), is a fair condition but the antenna and legs are rather shriveled and distorted.

Paratypes. 6 females [2 in USNM, 4 in UCRC] and 2 pupae [UCRC] were remounted from the same slide as the holotype, with the same label data. Additionally, 130 females [1 in AMUZ, 2 in BMNH, 2 in CNCI, 1 in EMEC, 2 in TAMU, 1 in UCDC, 15 in USNM, remainder in UCRC] as well as 10 pupae [UCRC] were remounted from the 17 original paratype slides. Their label data agree with the data published by DeBach (1979).

COMMENTS. This species is thelytokous, unlike the other species in the *lingnanensis* species group. A black streak on the posterior margin of scutellum is one of the distinguishing features of *A. riyadhi*. However, this black marking seems to be variable. The mesosomal setae of *A. riyadhi* are distinctly darker (dark and coarse) than in the other species of the "*holoxanthus-melinus* complex" of *Aphytis*. Other than that, this thelytokous species is morphologically indistinguishable from several other described species in the "*holoxanthus-melinus* complex" of the *lingnanensis* species group, such as *A. fisheri*, *A. holoxanthus*, *A. melinus*, and *A. yasumatsui* Azim.

Aphytis rolaspidis DeBach and Rosen

(related to *mytilaspidis* species group)

Aphytis flavus Quednau 1964: 102, 104-105.

Aphytis rolaspidis DeBach and Rosen 1976: 544 (replacement name for *A. flavus* Quednau [*nec* Nees, 1834]); Rosen and DeBach 1979: 497-499.

TYPE MATERIAL. Before remounting, there were no type specimens of this species in UCRC. One original Quednau's syntype slide with 4 female and 2 male specimens was received on loan from SANC, containing the lectotype female, 3 paralectotype females and 2 paralectotype males (one of which was invalidly designated by Rosen and DeBach in 1979 as an "allotype"). This contradicts the statement by Rosen and DeBach (1979) that this species was redescribed (p. 498) from "the entire syntype series: 5 females and 1 male... mounted on one slide". The awful condition of the original slide (the Hoyer's mountant under the coverslip was almost completely dark so that the specimens were scarcely visible even with good illumination) probably contributed to their mistake. Moreover, according to Quednau (1964), the syntype series of his *A. flavus* consisted of 5 females and 3 males, so apparently from the beginning there was some confusion about the number of the type specimens, all of which were mounted on the same slide. Rosen and

DeBach (1979) mentioned that the "additional" two males on that slide apparently belong to an undescribed species of *Aphytis*.

Lectotype female, designated by Rosen and DeBach (1979: 498). Original labels on the original syntype slide: 1. (red) "*Aphytis flavus* Quednau ♂, ♀ Syntypes F. W. Quednau det."; 2. "T 1096 *Rolaspis chaetachmae Ch. aristata* Durban (Nov.) [in pencil] 11.1962 J. Munting coll."; 3. (on the bottom of the slide) "*Aphytis ignotus* Comp. ♂, ♀ W. Quednau det."; 4. (on the bottom of the slide) "*Aphytis africanus* [crossed out in pencil] ♂, ♀ W. Quednau det."; 5. (on the bottom of the slide) "*Aphytis rolaspidis.*, n, name LECTOTYPE [in red]. ♀ + 2 ♂♂ ?another sp.? Det. DR and DeB 1973". After remounting, the lectotype female is labeled as follows: 1. "*Aphytis flavus* Quednau Syntype female *Aphytis rolaspidis* DeBach and Rosen Lectotype female (designated by DeBach and Rosen, 1976, see Rosen and DeBach 1979; labeled by J.-W. Kim and S. V. Triapitsyn 2002) W. Quednau, det."; 2. "South Africa, Durban, Natal, xi.1962, J. Munting (T 1096). Host: *Rolaspis chaetachmae* (Brain) on *Chaetachme aristata*. Remounted in Canada balsam by V. V. Berezovskiy at UCR". The lectotype, deposited in SANC, is complete and in fairly good condition.

Paralectotypes. 3 females and 2 males on separate slides, bearing the same label data as lectotype. All deposited in SANC except for 1 female that was retained in UCRC.

Aphytis roseni DeBach and Gordh

(unassigned to species group)

Aphytis roseni DeBach and Gordh 1974: 260-265; Rosen and DeBach 1979: 678-681.

TYPE MATERIAL. Before remounting, there were about 55 paratype specimens in UCRC on 11 slides, all mounted in a completely dry Hoyer's medium. In addition, the holotype female and the allotype male, mounted on the same slide in dry Hoyer's, were received from USNM. According to DeBach and Gordh (1974), this species was described from 84 specimens from Peru, Kenya, and Uganda (listed under "Material examined", including the female holotype, the male allotype, as well as 59 female and 13 male paratypes) but apparently the African specimens were not included in the type series (at least the specimens from Kenya in UCRC are not marked as paratypes). Rosen and DeBach (1979) redescribed this species from the type specimens from Peru only: the female holotype, the male allotype, as well as 48 female and 14 male paratypes from two different locations, so an obvious confusion exists regarding the true number of paratypes of this species.

Holotype female. Original labels: 1. (in black ink) "NAME *Aphytis* #2 [in pencil] *roseni* DeBach and Gordh HOLOTYPE ♀ ALLOTYPE ♂ DET. DeB. and G. 1973 COLL. O. Beingolea NO. VID.1:1 [DeBach code, in pencil] Div. Biol. Cont. Univ. Calif."; 2. (in black ink) "LOC. Peru, 90 mi. N. Lima, Huaru Valley DATE 8/V 1973 HOST *Selenaspidis articulatus* DET. Beingolea 1973 ON Citrus LTR. DTD. 11/V/1973". After remounting, the holotype is labeled as follows: 1. "*Aphytis roseni* ♀ DeBach and Gordh <u>HOLOTYPE</u> Remounted from Hoyer's into Canada balsam 2000 By M. Planoutene (UCR)"; 2. "Peru, 90 mi. N. of Lima, Huaru Valley, 8.v.1973, O. Beingolea. Host: *Selenaspidus articulatus* (Morgan) on citrus. DeBach code VID.1:1. Note on orig. slide: "LTR. DTD. 11/V/1973"; 3. (bar code) "UCRC ENT 003936". The holotype, deposited in USNM (on permanent loan from UCRC), is in moderately good condition, with the wings slightly damaged.

Paratypes. The allotype male [USNM], remounted from the same slide as the holotype (with the same label data). Other paratypes: 40 females and 8 males [1 female in EMEC; 1 female in CNCI; 1 female in TAMU; 1 female, 1 male in USNM;

remainder in UCRC], remounted from the original paratype slides, all with the same label data as holotype (det. by G. Gordh in 1973). Also 19 females and 5 males [UCRC], all labeled as follows: 1. "*Aphytis roseni* DeBach and Gordh PARATYPE Det. Gordh 1973 Remounted from Hoyer's into Canada balsam 2000 By M. Planoutene (UCR)"; 2. "Peru, Lima, 4.viii.1972, Oscar Beingolea. Host: *Selenaspidus articulatus* (Morgan) on citrus [DeBach codes follow, some with the following note from the original slide: "LAB REARED"].

NON-TYPE MATERIAL. Also remounted from Hoyer's into Canada balsam were 29 non-type specimens of *A. roseni* from Kenya.

Aphytis salvadorensis Rosen and DeBach
(unassigned to species group)

Aphytis salvadorensis Rosen and DeBach 1979: 696-699.

TYPE MATERIAL. This species was described from 14 specimens: the female holotype, the male allotype, as well as 5 female and 7 male paratypes (Rosen and DeBach 1979). Before remounting, there were 4 slides with about 7 paratypes in UCRC, all mounted in dry Hoyer's medium. The three other slides with the holotype (one slide) and the allotype (two slides, one of them with a forewing mounted separately) of *A. salvadorensis* were received from USNM. The entire type series had been originally mounted (apparently uncleared) in Canada balsam, from where it was remounted into Hoyer's (Rosen and DeBach 1979).

Holotype female. Original labels [In red ink]: 1. "NAME *Aphytis salvadorensis*, n. sp. HOLOTYPE ♀ DET. DR 1975 COLL. P. A. Berry NO. VID.14:1 [DeBach code, in pencil] Div. Biol. Cont. Univ. Calif."; 2. "LOC. San Miguel, El Salvador DATE V-30, 1957 HOST ? "scale" DET. 19 Remounted, balsam - [to] Hoyer".

After remounting, the holotype is labeled as follows: 1. "*Aphytis salvadorensis* ♀

Rosen and DeBach <u>HOLOTYPE</u> Det. D. Rosen 1975 Remounted from Hoyer's into

Canada balsam 2000 By M. Planoutene (UCR)"; 2. "El Salvador, San Miguel,

30.v.1957, P. A. Berry. Host: undetermined scale insect. Note on orig. slide

"remounted from Balsam to Hoyer" DeBach code VID.14:1"; 3. (bar code) "UCRC

ENT 004362". The holotype, deposited in USNM (on permanent loan from UCRC),

is in fair condition but incomplete (flagellum of one antenna and parts of one foreleg

are missing); the head, a hind wing, and two legs (middle and hind) are detached

from the body.

Paratypes. The allotype male on two slides (one forewing, mounted separately, is

still in Hoyer's) [USNM], 5 female and 6 male paratypes on separate slides (some

have missing body parts) [1 female and 1 male in USNM, remainder in UCRC], with

the same label data as holotype.

Aphytis sensorius DeBach and Rosen

(*chrysomphali* species group)

Aphytis sensorius DeBach and Rosen 1976: 544-545; Rosen and DeBach 1979: 620-

622.

TYPE MATERIAL. Before remounting, there were 44 slides of *A. sensorius* in

UCRC that contained about 30 paratype specimens, all mounted in Hoyer's. An

additional slide, containing the female holotype and the male allotype of this species,

was received from USNM.

Holotype female. Original labels: 1. [in red and black ink] "NAME *Aphytis*

sensorius, n. sp. HOLOTYPE ♀ ALLOTYPE ♂ DET. DR 1968 NO. XVI-66

COLL. M. A. Ghani Dept. Biol. Cont. Univ. Calif."; 2. [In black ink] "LOC. Murree

Hills, Rawalpindi, Pakistan DATE 23 Aug. 1966 HOST *Leucaspis coniferarum* DET 19 ON *Pinus roxburghii* VIB.24:4 [DeBach's code, in pencil]". After remounting, the holotype female is labeled as follows: 1. "*Aphytis sensorius* ♀ DeBach and Rosen Holotype Remounted from Hoyer's into Canada balsam 1999 By M. Planoutene (UCR)"; 2. "Pakistan, Rawalpindi, Murree Hills, 23.viii.1966, M. A. Ghani. Host: *Anamaspis coniferarus* (Hall and Williams) on *Pinus roxburghii*. No. on orig. slide "XVI-66". DeBach code VIB.24:1"; 3. (bar code) "UCRC ENT 004921". The holotype, deposited in USNM (on permanent loan from UCRC), is in fair condition but missing flagellum of one antenna; the head and the other flagellum are detached.

Paratypes. The allotype male on slide [USNM], remounted from the same original slide as the holotype; also 25 female and 23 male paratypes [1 female and 1 male in EMEC; 1 female and 1 male in CNCI; 1 female and 1 male in TAMU; 1 female, 1 male in USNM; 1 female, 1 male in ZIN; remainder in UCRC], remounted from the 4 original slides in UCRC, with the same label data as holotype; many of these are incomplete (missing some body parts).

Aphytis setosus DeBach and Rosen
(unassigned to species group)

Aphytis ciliatus Quednau 1964: 102.

Aphytis setosus DeBach and Rosen 1976: 545 (replacement name for *A. ciliatus* Quednau, 1964 [*nec* Dodd, 1917]); Rosen and DeBach 1979: 687-690.

TYPE MATERIAL. Before remounting, there were no type specimens of this species in UCRC. A slide with the holotype male of *A. ciliatus* Quednau (mounted in a dark Hoyer's medium under the same coverslip with several non-type specimens of *A. funicularis* Compere) was received on loan from SANC. The "allotype" female of *A. setosus* was invalidly designated by Rosen and DeBach (1979) from unrelated, non-type material (the species was originally described from a single male holotype); the slide containing that female and several other non-type specimens collected in 1970 in South Africa was returned to SANC long after Rosen and DeBach (1979) had mistakenly indicated UCRC as their depository.

Holotype male. Original labels: 1. (in black ink) "*Aphytis funicularis* Compere *A. ciliatus* n. sp. ♂, ♀ Type W. Quednau det. [bottom half of the label marked in red]"; 2. (in black ink) "T 1096 *Rolaspis chaetachmae Ch. aristata* Durban 11.1962 J. Munting coll.". After remounting, the holotype is labeled as follows: 1. "*Aphytis ciliatus* Quednau Holotype male *Aphytis setosus* DeBach and Rosen Holotype male (designated by DeBach and Rosen, 1976, see Rosen and DeBach 1979; labeled by J.-W. Kim and S. V. Triapitsyn 2002)"; 2. "South Africa, Durban, Natal, xi.1962, J. Munting (T 1096). Host: *Rolaspis chaetachmae* (Brain) on *Chaetachme aristata*. Remounted in Canada balsam by V. V. Berezovskiy at UCR". The holotype, deposited in SANC, is in very good condition, with flagellum of one antenna detached.

Aphytis simmondsiae DeBach
(*lingnanensis* species group)

Aphytis simmondsiae DeBach 1984: 103-112.

TYPE MATERIAL. Before remounting, there were 38 paratype specimens in UCRC on 19 slides, all mounted in dry Hoyer's. The holotype of this species, mounted in Hoyer's, is not in UCRC, as specified by DeBach (1984), but is currently deposited in USNM under type number 103547.

Paratypes. 38 females [1 in CNCI, 1 in EMEC, 1 in UCDC, 2 in TAMU, 7 in USNM, remainder in UCRC]. The label data agree with the data provided in the original description (DeBach 1984).

COMMENTS. According to DeBach (1984), the type series of *A. simmondsiae* consisted of numerous females and males on 40 slides, some of which were supposedly distributed among the taxonomic collections listed in the original description.

Aphytis taylori Quednau
(related to *mytilaspidis* species group)

Aphytis taylori Quednau 1964: 104-106; Rosen and DeBach 1979: 494-496.

TYPE MATERIAL. According to the original description, this species was described from the series of 40 female and 7 male syntypes mounted on 5 slides (Quednau 1964). Before remounting, there was 1 syntype slide of this species in UCRC, and the remaining 4 syntype slides were received on loan from SANC. All the original slides had the Faure solution mountant under the coverslips almost completely dark so that the specimens were scarcely visible even with good illumination; moreover, as noted by Rosen and DeBach (1979), the specimens were in rather poor state (no complete specimens could be found).

Lectotype female. After remounting, a lectotype female, here designated, was chosen from one of the SANC syntypes to ensure applicability of the name to the proper specimens and because the mounting medium and labeling of the type specimens of this species have been changed.

Original labels on the lectotype slide: 1. (red) "*Aphytis taylori* Quednau ♀ Syntypes F. W. Quednau det."; 2. "*Aspidiotus capensis* on *Cycas* T744 T840 Port Elisabeth S.A. 2.62 J. S. Taylor 3 [circled in pencil]" 3. (on the underside) "*Aphytis africanus* V W. Quednau det."; 4. (on the underside) "*Aphytis mytilaspidis* (Le Baron) ♀ W. Quednau det.". After remounting, the lectotype female is labeled as follows: 1. "*Aphytis taylori* Quednau ♀ Syntype (labeled by J.-W. Kim and S. V. Triapitsyn 2002)"; 2. "*Aphytis taylori* Quednau, 1964 LECTOTYPE ♀ Des. S. Triapitsyn 2004"; 3. "South Africa, Port Elisabeth, ii.1962, J. S. Taylor. Host: *Aspidiotus capensis* Newstead on *Cycas* sp. Remounted in Canada balsam by V. V. Berezovskiy at UCR". The lectotype, deposited in SANC, lacks the flagellum of one antenna, one middle leg, and most of one hind wing, but otherwise is in fair condition.

Paralectotypes. 9 females [8 in SANC, 1 in UCRC], remounted from the lectotype slide, with the same label data. 1 female and 1 male [UCRC], remounted from the original syntype slide, also with the same label data. 8 females and 2 males [SANC], remounted from the original syntype slide, with the same label data except the original labels did not have collector's code numbers and did not mention presence of the males on this slide. 10 females [SANC], remounted from the original syntype slide, with the same label data as lectotype. 5 females and 2 males [SANC], remounted from the original syntype slide, also with the same label data as lectotype. Altogether, remounted were 34 out of 40 syntype females and 5 out of 7 syntype males mentioned by Quednau (1964); because we remounted all the specimens from all 5 original slides, it appears that F. W. Quednau miscounted the number of specimens on his original 5 syntype slides.

Aphytis theae (Cameron)

(*funicularis* species group)

Aphelinus theae Cameron 1891: 183-184.

Aphytis theae (Cameron): Rosen and DeBach 1979: 656-661.

TYPE MATERIAL. A neotype for this species was designated by Rosen and DeBach (1979). The original slide from the "neotype series", containing the female neotype and the male "allotype" [an invalid designation by Rosen and DeBach (1979)] as well as four other specimens, was received from USNM.

Neotype female. Original labels: 1. [in black ink] "NAME *Aphytis theae* (Cameron) lowest ♀ = NEOTYPE middle ♂ = ALLOTYPE DET. DR 1976 COLL. F. Collins NO. see letters Div. Biol. Cont Univ. Calif."; 2. [In black ink] "Honey Plant Greenhouse LOC. Univ. of Florida, Gainesville, Florida DATE V -27 1976 HOST *Fiorinia theae* (orig. ex. India) DET 19 ON *Camellia japonica* VIC.21:1 [DeBach's code, in pencil]". After remounting, the neotype female is labeled as follows: 1. "*Aphytis theae* ♀ (Cameron) Neotype Det. D. Rosen 1976 Remounted from Hoyer's into Canada balsam 2000 By M. Planoutene (UCR)"; 2. "U.S.A., Florida, Gainesville, USDA Biological Control Laboratory (Insectary), 27.v.1976, F. Collins. Host: *Fiorinia theae* Green on *Camellia japonica* (tea plants). Orig. India, Jorhat (Assam). DeBach code VIC.21:1"; 3. (bar code) "UCRC ENT 003701". The neotype, deposited in USNM (on permanent loan from UCRC), is complete and in fair condition.

NON-TYPE MATERIAL. The invalidly designated "allotype" male [USNM], remounted from the same original slide as the neotype, along with 2 females and 2 males from the "neotype series", remounted from the same slide [UCRC]; also 6 females and 6 males, remounted from 2 original slides, females from one slide and males from the other [1 female and 1 male in USNM, remaining specimens in in UCRC], with the same label data (originally labeled as "neotype series").

Also remounted from Hoyer's into Canada balsam were additional 12 non-type specimens of *A. theae* from India.

Aphytis tucumani Rosen and DeBach
(related to *proclia* species group)

Aphytis tucumani Rosen and DeBach 1979: 416-419.

TYPE MATERIAL. Before remounting, there was 1 slide with 9 paratype males in UCRC, mounted in Hoyer's. These males were not remounted. A slide with the holotype female, the allotype male, and 11 paratypes (4 females and 7 males) of *A. tucumani* was received from USNM.

Holotype female. Original labels [In black ink]: "NAME *Aphytis tucumani*, n. sp. uppermost ♀ = HOLOTYPE up/right ♂ = ALLOTYPE DET. DR 1976 COLL. Fidalgo, P. NO. Div. Biol. Cont. Univ. Calif. PARATYPES"; 2. "LOC. Argentina, Tucumán, S. M. Tucuman DATE 13.VI 1972 HOST *Diaspis ?echinocacti* DET. 19 ON *Cereus* sp. IID.6:2 [DeBach code, in pencil] ". After remounting, the holotype is labeled as follows: 1. "*Aphytis tucumani* ♀ Rosen and DeBach <u>HOLOTYPE</u> Det. D. Rosen 1976 Remounted from Hoyer's into Canada balsam 2000 By M. Planoutene (UCR)"; 2. "Argentina, Tucumán, S. M. de Tucumán, 13.vi.1972, P. Fidalgo. Host: *Diaspis ?echinocacti* (Bouché) on *Cereus* sp. DeBach code IID.6:2"; 3. (bar code) "UCRC ENT 004291". The holotype, deposited in USNM (on permanent loan from UCRC), is in good condition but incomplete (lacking both hind wings); one forewing is detached from the body.

Paratypes. The allotype male [USNM], 4 females and 7 males on separate slides [1 female and 1 male in IMLA; 1 female and 1 male in MLPA, 1 male in USNM, remainder in UCRC], remounted from the same slide as the holotype. Also 9 males [UCRC], mounted in Hoyer's on a slide under the same coverslip, labeled: "NAME *Aphytis tucumani*, n. sp. PARAtype ♂♂ DET. DR 1976 COLL. Fidalgo, P. NO. Div. Biol. Cont. Univ. Calif."; 2. "LOC. CHACO, 20 km W R. Saenz Peña,

Argentina DATE 19.VIII 1972 HOST *Diaspis ?echinocacti* DET. 19 ON *Cereus* sp. IID.6:1 [DeBach code, in pencil]".

Aphytis vandenboschi DeBach and Rosen
(*proclia* species group)

Aphytis vandenboschi DeBach and Rosen 1976: 543; Rosen and DeBach 1979: 400-402.

TYPE MATERIAL. This uniparental species was described from the holotype female and numerous female paratypes, which had never been counted. Before remounting, there were 9 slides of *A. vandenboschi* in UCRC that contained several hundred paratype specimens, all mounted in Hoyer's. Another slide, containing the female holotype and 20 female paratypes of this species, was received from USNM.

Holotype female. Original labels: 1. [in red and blue ink] "n. sp. [in pencil] NAME *Aphytis vandenboschi* n. sp. HOLOTYPE ♀ PARATYPEs DET. BY DR 1969 S and R 64-A-33 LOT NO. JA Coll. Van den Bosch Dept Biol. Conn. Univ Calif"; 2. [In blue ink] "Rec'd from Albany LOC. Kashiwubaru, Nr. Fukuoka, Japan. DATED May 20 1964 HOST San Jose Scale IIB.23:1 [DeBach's code, in pencil] DET. BY 19 ON Pear Trees COLL. (Glenn Finney)". After remounting, the holotype female is labeled as follows: 1. "*Aphytis vandenboschi* ♀ DeBach and Rosen HOLOTYPE Remounted from Hoyer's into Canada balsam 2000 By M. Planoutene (UCR)"; 2. "U.S.A., Ca., Mat. Received in UCR Quarantine (S&R 64-A-33 from Albany lab (UC Berkeley), reared there by G. Finney on *Aspidiotus nerii* (Bouché). Orig. from Japan, Kashiwubaru, nr. Fukuoka, 20.v.1964, R. van den Bosch, ex. *Quadraspidiotus perniciosus* (Comstock) on pear. DeBach code IIB.23:1"; 3. (bar code) "UCRC ENT 004801". The holotype, deposited in USNM (on permanent loan from UCRC), is in excellent condition but missing a tibia and tarsus of one foreleg; one forewing is detached from the body.

Paratypes. 666 females on individual slides, remounted from 10 original slides (including the 20 specimens from the holotype slide) [2 females each in AMUZ, ANIC, CNCI, EMEC, IMLA, SANC, TAMU, UCDC, USNM, ZIN; remainder in UCRC], with the same label data as holotype; many of them also have the date they were cultured at UCR quarantine (v.1965).

NON-TYPE MATERIAL. Also remounted from Hoyer's into Canada balsam were additional 20 non-type specimens of *A. vandenboschi* from Japan, including 1 pupa from one of the paratype slides.

Aphytis vittatus (Compere)

(*vittatus* species group)

Paraphytis vittatus Compere 1925: 129-133.
Aphytis vittatus (Compere): Rosen and DeBach 1979: 236-238.

TYPE MATERIAL. The holotype female and the allotype male of this species are in USNM. There are 2 female (one of them incomplete) paratypes of *A. vittatus* on separate slides and also 1 male paratype on 2 slides in UCRC, all mounted in Canada balsam, as follows: 1 female, labeled: 1. "♀ *Paraphytis vittata* Compere = *Marietta* PARATYPE - "KOH" Ex *Lepidosaphes tubulorum* Ferris Silvestri Shipment No. A48. Amoy, China Jan. 24, 1925 *Aphytis vittatus* [in pencil]"; 2. "Ex. *Lepidosaphes tubulorum* Ferris. Amoy, China. Silvestri shipment No. A48. Jan. 24, 1925 IA.1:2 [DeBach code, in pencil] U. C. Cit. Exp. Sta.". 1 female (incomplete), labeled: 1. "*Marietta* sp. *vittata* [in pencil] Amoy, China - F. Silvestri IA.1:4 [DeBach code, in pencil] Ex *Lepidosaphes* (good wings) U. C. Cit. Exp. Sta.". 1

male, labeled: 1. "♂ *Paraphytis vittata* Compere = *Marietta* <u>PARATYPE</u> Boiled in KOH Ex *Lepidosaphes tubulorum* Ferris Silvestri shipment No. A48 From Amoy, China Jan. 24, 1925."; 2. "Ex. *Lepidosaphes tubulorum* Ferris. Amoy, China. Silvestri shipment No. A48. Jan. 24, 1925 IA.1:1 [DeBach code, in pencil] U. C. Cit. Exp. Sta.", with a forewing mounted on a separate slide, labeled: 1. "*Paraphytis vittata* H. Compere New genus and species Wing - body on another slide. PARATYPE U. C. Cit. Exp. Sta"; 2. "Ex *Lepidosaphes tubulorum* Ferris. Amoy, China. Silvestri shipment No. A48 Jan. 24, 1925 IA.1:3 [DeBach code, in pencil] U. C. Cit. Exp. Sta".

Aphytis yanonensis DeBach and Rosen

(*lingnanensis* species group)

Aphytis yanonensis DeBach and Rosen 1982: 628-633.

TYPE MATERIAL. The female holotype of this species was deposited in IZCAS (DeBach and Rosen 1982). The following paratype specimens, all but one mounted in good quality Hoyer's (ringed with Glyptal®), were received from the estate of the late Mike Rose after the remounting project was over.

Paratypes [UCRC]. 9 females on individual slides, labeled: 1. "Loc UCR culture orig R81-42 Date X-5 1981 Host *Aspidiotus hederae* Det UCR cult. 1981 On lemon fruit"; 2. "Name *Aphytis yanonensis* ♀ DeBach + Rosen PARATYPE Coll Rose No. R81-42"; 3. (green) "This ♀ utilized in xing test w/ CI ♂ *A. lingnanensis*". 4 females on individual slides, labeled: 1. "Loc Shizuoka Exp. St., Shizuoka, Japan Date III 25 1981 Host *Unaspis yanonensis* On laboratory culture orig. China"; 2. "Name *Aphytis* ♀ *yanonensis* DeBach & Rosen PARATYPE Coll K. Furuhashi No.

see letter III.25.81". 5 females on 3 slides, labeled: 1. "Loc orig. Shizuoka, Japan (lab cult) Date IX-16 1981 Host *Aspidiotus hederaer* On lemon fruit F1 newly emerged"; 2. "Name *Aphytis* ♀ *yanonensis* DeBach & Rosen PARATYPE Coll Rose No. s No. R81-42". 11 females on slide, labeled: 1. "Loc orig. Shizuoka, Japan F2 lab SCFS Date X-5 1981 Host *Aspidiotus nerii* On lemon fruit"; 2. "Name *Aphytis yanonensis* DeBach & Rosen PARATYPES ♀ ♀ Coll Ro".

Aphytis yasumatsui Azim

(*lingnanensis* species group)

Aphytis yasumatsui Azim 1963: 284-287; Rosen and DeBach 1979: 558-561.

TYPE MATERIAL. Before remounting, there were no syntype specimens of this species in UCRC. One syntype (on a slide in Hoyer's) was then received from USNM. According to Rosen and DeBach (1979), the rest of the syntype specimens from the type series of *A. yasumatsui* are in the collection of the Entomological Laboratory, Kyushu University, Fukuoka, Japan.

Syntype male [USNM]. Original labels: 1. "*Aphytis yasumatsui*, Azim 1963, inconsistent with original description ["co-type" is written on an additional green label]; 2. "Hakozobi, Fukuoka, Kyushu, Japan, Sept. 10, 1962. Host: *Chrysomphalus bifasciculatus*, Det. by Azim, VB.12:1, Coll. A. Azim". After remounting, the syntype is labeled as follows: 1. "*Aphytis yasumatsui* ♂ Azim, SYNTYPE, Remounted from Hoyer's into Canada balsam 2000 By M. Planoutene (UCR)"; 2. "Japan, Kyushu, Fukuoka, Hakozaki, 10.ix.1962, A. Azim. Host: *Chrysomphalus bifasciculatus* Ferris, "co-type; inconsistent with description". DeBach code VB.12:1"; 3. (bar code) "UCRC ENT 004290".

NON-TYPE MATERIAL. Also remounted were 124 non-type specimens of *A. yasumatsui* from Japan [UCRC]. Among them, 1 female and 3 males may belong to the original material of this species although no indication of its possible syntype status could be found. These specimens are now labeled as follows: "Japan, Kyushu, Fukuoka, 1-3.vi.1961, A. Azim. Host: *Chrysomphalus bifasciculatus* Ferris on *Illicium religiosum* Note on orig. slide "D-letter of July 17" DeBach code VB.12:2". Also present in UCRC is an unremounted Hoyer's slide with a pupa, likely from the same original material, labeled: 1. "Pupa of *Aphytis* "D" *Chrysomphalus* sp. infested on *Illicium religiosum*" (possibly in A. Azim's handwriting); "See adult slide VB.12:3" [added later in pencil]; 2. "See slide of Adults *Aphytis yasumatsui* Azim, 1963 Young pupa see letter of July 17, 1961 from A. Azim".

Appendix

Depositories of the primary types of the nominal species of *Aphytis* treated in this catalog

Aphytis spp.	Type status	Present depository	Comments
A. acrenulatus DeBach and Rosen	holotype	USNM	on loan from UCRC
A. acutaspidis Rosen and DeBach	holotype	USNM	on loan from UCRC
A. africanus Quednau	lectotype	USNM	on loan from UCRC
A. amazonensis Rosen and DeBach	holotype	USNM	on loan from UCRC
A. anneckei DeBach and Rosen	holotype	USNM	on loan from UCRC
A. anomalus Compere	holotype	USNM	
A. antennalis Rosen and DeBach	holotype	USNM	on loan from UCRC
A. australiensis DeBach and Rosen	holotype	USNM	on loan from UCRC
A. bangalorenis Rosen and DeBach	holotype	?lost	never deposited to UCRC
A. bedfordi Rosen and DeBach	holotype	SANC	returned from UCRC
A. capensis DeBach and Rosen	holotype	USNM	on loan from UCRC
Aphelinus capitis Rust	holotype	USNM	
A. cercinus Compere	holotype	USNM	
A. citrinus Compere	lectotype	USNM	on loan from UCRC
A. cochereaui DeBach and Rosen	holotype	USNM	on loan from UCRC
A. coheni DeBach	lectotype	USNM	on loan from UCRC
A. comperei DeBach and Rosen	holotype	USNM	on loan from UCRC
A. confusus DeBach and Rosen	holotype	SANC	returned from UCRC
A. cylindratus Compere	lectotype	USNM	on loan from UCRC
A. dealbatus Compere	holotype	USNM	

Depositories of the primary types of the nominal species of *Aphytis* treated in this catalog continued

Aphytis spp.	Type status	Present depository	Comments
A. debachi Azim	syntype	UCRC	
A. desantisi DeBach and Rosen	holotype	USNM	on loan from UCRC
A. equatorialis Rosen and DeBach	lectotype	USNM	on loan from UCRC
A. fabresi DeBach and Rosen	holotype	USNM	on loan from UCRC
A. fisheri DeBach	lectotype	USNM	on loan from UCRC
A. funicularis Compere	holotype	USNM	
A. gordoni DeBach and Rosen	holotype	USNM	on loan from UCRC
A. griseus Quednau	lectotype	USNM	on loan from UCRC
A. hispanicus (Mercet)	holotype	MNMS	returned from UCRC
A. holoxanthus DeBach	lectotype	USNM	on loan from UCRC
A. hyalinipennis Rosen and DeBach	holotype	USNM	on loan from UCRC
A. ignotus Compere	holotype	?lost	not found in USNM
A. immaculatus Compere	lectotype	USNM	on loan from UCRC
A. japonicus DeBach and Azim	lectotype	USNM	on loan from UCRC
A. landii Rosen and DeBach	holotype	?lost	never deposited to UCRC
A. lepidosaphes Compere	lectotype	USNM	on loan from UCRC
A. limonus (Rust)	holotype	USNM	
A. lingnanensis Compere	lectotype	USNM	on loan from UCRC
A. longicaudus Rosen and DeBach	holotype	USNM	on loan from UCRC
A. luteus (Ratzeburg)	lectotype	UCRC	to be returned to DEI
A. malayensis Rosen and DeBach	holotype	USNM	on loan from UCRC
A. mandalayensis Rosen and DeBach	holotype	USNM	on loan from UCRC

Depositories of the primary types of the nominal species of *Aphytis* treated in this catalog continued

Aphytis spp.	Type status	Present depository	Comments
A. margaretae DeBach and Rosen	holotype	USNM	on loan from UCRC
A. mazalae DeBach and Rosen	holotype	USNM	on loan from UCRC
A. melanostictus Compere	holotype	?missing	likely not marked in UCRC
A. melinus DeBach	lectotype	USNM	on loan from UCRC
A. merceti Compere	holotype	USNM	
A. mimosae DeBach and Rosen	holotype	SANC	returned from UCRC
A. mytilaspidis (Le Baron)	neotype	USNM	on loan from UCRC
A. nigripes (Compere)	holotype	USNM	
A. obscurus DeBach and Rosen	holotype	USNM	on loan from UCRC
A. paramaculicornis DeBach and Rosen	holotype	USNM	on loan from UCRC
A. perplexus Rosen and DeBach	holotype	USNM	on loan from UCRC
A. philippinensis DeBach and Rosen	holotype	USNM	on loan from UCRC
A. phoenicis DeBach and Rosen	holotype	USNM	on loan from UCRC
A. pilosus DeBach and Rosen	holotype	SANC	returned from UCRC
A. pinnaspidis Rosen and DeBach	holotype	USNM	on loan from UCRC
Aphelinus quaylei Rust	holotype	USNM	
A. riyadhi DeBach	holotype	USNM	on loan from UCRC
A. rolaspidis DeBach and Rosen	lectotype	SANC	
A. roseni DeBach and Gordh	holotype	USNM	on loan from UCRC
A. salvadorensis Rosen and DeBach	holotype	USNM	on loan from UCRC
A. sankarani Rosen and DeBach	holotype	?lost	never deposited to UCRC

Depositories of the primary types of the nominal species of *Aphytis* treated in this catalog continued

Aphytis spp.	Type status	Present depository	Comments
A. sensorius DeBach and Rosen	holotype	USNM	on loan from UCRC
A. setosus DeBach and Rosen	holotype	SANC	
A. simmondsiae DeBach	holotype	USNM	donated from UCRC
A. spiniferus Compere and Annecke	holotype	USNM	
A. taylori Quednau	lectotype	SANC	
A. theae (Cameron)	neotype	USNM	on loan from UCRC
A. tucumani Rosen and DeBach	holotype	USNM	on loan from UCRC
A. vandenboschi DeBach and Rosen	holotype	USNM	on loan from UCRC
A. vittatus (Compere)	holotype	USNM	
A. yanonensis DeBach and Rosen	holotype	IZCAS	
A. yasumatsui Azim	syntype	USNM	possibly others in UCRC

Literature Cited

Annecke, D.P. 1963. New and little known species of South African Aphelinidae (Hymenoptera: Aphelinidae). Annals and Magazine of Natural History (Series 13) 6:337-352.

Azim, A. 1963. Systematic and biological studies on the genus *Aphytis* Howard (Hymenoptera, Aphelinidae) of Japan Part 1. Taxonomy. Journal of the Faculty of Agriculture, Kyushu University 12:265-290.

Bouček, Z. 1964. Proctotrupoidea und Chalcidoidea aus den Resten der Ratzeburg-Sammlung. Beiträge zur Entomologie 14:663-673.

Cameron, P. 1891. Hymenopterological notices. Memoirs and Proceedings of the Manchester Literary and Philosophical Society (4th Series) 4:182-194 + plate 1.

Compere, H. 1925. A new genus and species of Aphelinidae (Hymenoptera) from China. Transactions of the American Entomological Society 51:129-134.

_____. 1936. Notes on the classification of the Aphelinidae with descriptions of new species. University of California Publications in Entomology 6:277-321.

_____. 1955. A systematic study of the genus *Aphytis* Howard (Hymenoptera, Aphelinidae) with descriptions of new species. University of California Publications in Entomology 10:271-319.

Compere, H. and D.P. Annecke. 1961. Descriptions of parasitic Hymenoptera and comments (Hymenopt.: Aphelinidae, Encyrtidae, Eulophidae). Journal of the Entomological Society of South Africa 24:17-71.

DeBach, P. 1959. New species and strains of *Aphytis* (Hymenoptera, Eulophidae) parasitic on the California red scale, *Aonidiella aurantii* (Mask.), in the Orient. Annals of the Entomological Society of America 52:354-362.

_____. 1960. The importance of taxonomy to biological control as illustrated by the cryptic history of *Aphytis holoxanthus* n.sp. (Hymenoptera: Aphelinidae), a parasite of *Chrysomphalus aonidum*, *Aphytis coheni* n.sp., a parasite of *Aonidiella aurantii*. Annals of the Entomological Society of America 53:701-705.

_____. 1979. *Aphytis riyadhi* n.sp. (Hym.: Aphelinidae), a parasite of *Aonidiella* spp. (Hom.: Diaspididae). Entomophaga 24:133-138.

_____ 1984. *Aphytis simmondsiae* n. sp. (Hymenoptera: Aphelinidae), a parasite of jojoba scale, *Diaspis simmondsiae* (Homoptera: Diaspididae). Folia Entomológica Mexicana 60:103-112.

DeBach, P. and A. Azim. 1962. *Aphytis japonicus* n. sp., a parasite of *Chrysomphalus bifasciculatus* Ferris in Japan (Hymenoptera: Aphelinidae). Mushi 36:1-8.

DeBach, P. and G. Gordh. 1974. A new species of *Aphytis* that attacks important armored scale insects. Entomophaga 19:259-265.

DeBach, P. and D. Rosen. 1976. Twenty new species of *Aphytis* (Hymenoptera: Aphelinidae) with notes and new combinations. Annals of the Entomological Society of America 69:541-545.

_____. 1982. *Aphytis yanonensis* n. sp. (Hymenoptera, Aphelinidae), a parasite of *Unaspis yanonensis* (Kuwana) (Homoptera, Diaspididae). Kontyû 50:626-634.

Delucchi, V. 1964. Une nouvelle espèce d'*Aphytis* du groupe *chilensis* Howard (Hym., Chalcidoidea, Aphelinidae). Revue de Pathologie Végétale et d'Entomologie Agricole de France 43:135-140.

De Santis, L. 1948. Estudio monográfico de los afelínidos de la República Argentina (Hymenoptera, Chalcidoidea). Revista del Museo de La Plata (Nueva serie), Sección Zoología 5:23-280.

Dozier, H.L. 1933. Miscellaneous notes and descriptions of chalcidoid parasites (Hymenoptera). Proceedings of the Entomological Society of Washington 35:85-100.

Gahan, A.B. 1924. Some new parasitic Hymenoptera with notes on several described species. Proceedings of the United States National Museum 65 (Art. 4, No. 2517):1-23.

Gomes, J.G. 1942(1941). Subsídios à sistemática dos calcidídeos brasileiros. Boletim da Escola Nacional de Agronomia 2:1-37 + estampas 1-4.

Hayat, M. 1998. Aphelinidae of India (Hymenoptera: Chalcidoidea): A taxonomic revision. Memoirs on Entomology, International 13:i-viii + 1-416.

Howard, L.O. 1900. A new genus of Aphelininae from Chile. The Canadian Entomologist 32:167-168.

Jasnosh, V.A. and Myartseva, S.N. 1971. [Two new aphelinid species (Chalcidoidea, Aphelinidae) – parasites of armored scales (Coccoidea, Diaspididae) from Middle Asia]. Izvestiya Akademii Nauk Turkmenskoy SSR, Seriya Biologicheskikh Nauk 6:35-41.

Le Baron, W. 1870. The chalcideous parasite of the apple-tree bark louse (*Chalcis* [*Aphelinus*] *mytilaspidis*, n. sp.). The American Entomologist and Botanist 2:360-362.

Mercet, R.G. 1911. Los Calcídidos parásitos de Cóccidos. Boletín de la Real Sociedad Española de Historia Natural 11:507-515.

_____. 1912a. Un parásito del "poll-roig". Boletín de la Real Sociedad Española de Historia Natural 12:135-140.

_____. 1912b. Los enemigos de los parásitos de las plantas. Los Afelíninos. Trabajos del Museo Nacional de Ciencias Naturales, Serie Zoológica 10:1-306.

Nikol'skaya, M.N. and V.A. Jasnosh. 1966. [Aphelinids of the European Part of the USSR and the Caucasus]. Opredeliteli po faune SSSR, izdavaemye Zoologicheskim institutom Akademii Nauk SSSR, Vol. 91. Nauka, Moscow and Leningrad, 295 pp. (In Russian).

Platner, G.R., R.K. Velten, M. Planoutene and J.D. Pinto. 1999. Slide-mounting techniques for *Trichogramma* (Trichogrammatidae) and other minute parasitic Hymenoptera. Entomological News 110:56-64.

Quednau, F.W. 1964. A contribution on the genus *Aphytis* Howard in South Africa (Hymenoptera: Aphelinidae). Journal of the Entomological Society of South Africa 27:86-116.

Ratzeburg, J.T.C. 1852. Die Ichneumonen der Forstinsecten in forstlicher und entomologischer Beziehung ein Anhang zur Abbildung und Beschreibung der Forstinsecten, Vol. 3. Berlin, III-XVIII + 272 pp. + 3 plates.

Rosen, D. 1994. Fifteen years of *Aphytis* research - an update. *In*: Advances in the study of *Aphytis* (Hymenoptera: Aphelinidae). Intercept Ltd., Andover, pp. 1-9.

Rosen, D. and P. DeBach. 1979 (1978). Species of *Aphytis* of the world (Hymenoptera: Aphelinidae). Series Entomologica, Vol. 17, Israel University Press, Jerusalem, Dr. W. Junk Publishers, 801 pp.

———. 1986. Three new species of *Aphytis* (Hym.: Aphelinidae), parasites of *Pseudaulacaspis* spp. (Hom.: Diaspididae) in India and Australia. Entomophaga 31:139-151.

Rust, E.W. 1915. Three new species of *Aphelinus* (Hym.). Entomological News 26:73-77.

Schauff, M. 1985. Slide mounting of types. Transfer of *Aphytis* types (Collection notes). Chalcid Forum 5:4-5.

Shafee, S.A. 1970. New genus of Aphelinidae recorded from Ootacamund (India) (Hymenoptera). Mushi 43:143-147.

Taeger, A., H. Gaedike and S.M. Blank. 2005. Katalog der primären Hymenopteren-Typen des DEI: (unter Ausschluss der Symphyta und Sphecidae s.l.). Beiträge zur Entomologie 55:151-250.

Upton, M.S. 1993. Aqueous gum-chloral slide mounting media: an historical review. Bulletin of Entomological Research 83:267-274.

FIGURES

Figs 1-4. Major problems in UCR collection of slide-mounted *Aphytis* (prior to remounting). 1. Hoyer's breakdown on the paratype slide of *A. cochereaui* (numerous specimens under the same coverslip ringed with Zut®). 2. Hoyer's breakdown on the paratype slide of *A. fabresi* (the mounting medium is completely dry, the coverslip ringed with Zut®). 3. Hoyer's breakdown on the paratype slide of *A. acrenulatus* (the mounting medium is almost dry, the coverslip ringed with Zut®). 4. Hoyer's breakdown on the slide of *A. opuntiae* (the mounting medium under two coverslips ringed with Zut® broke down and not covering the specimens).

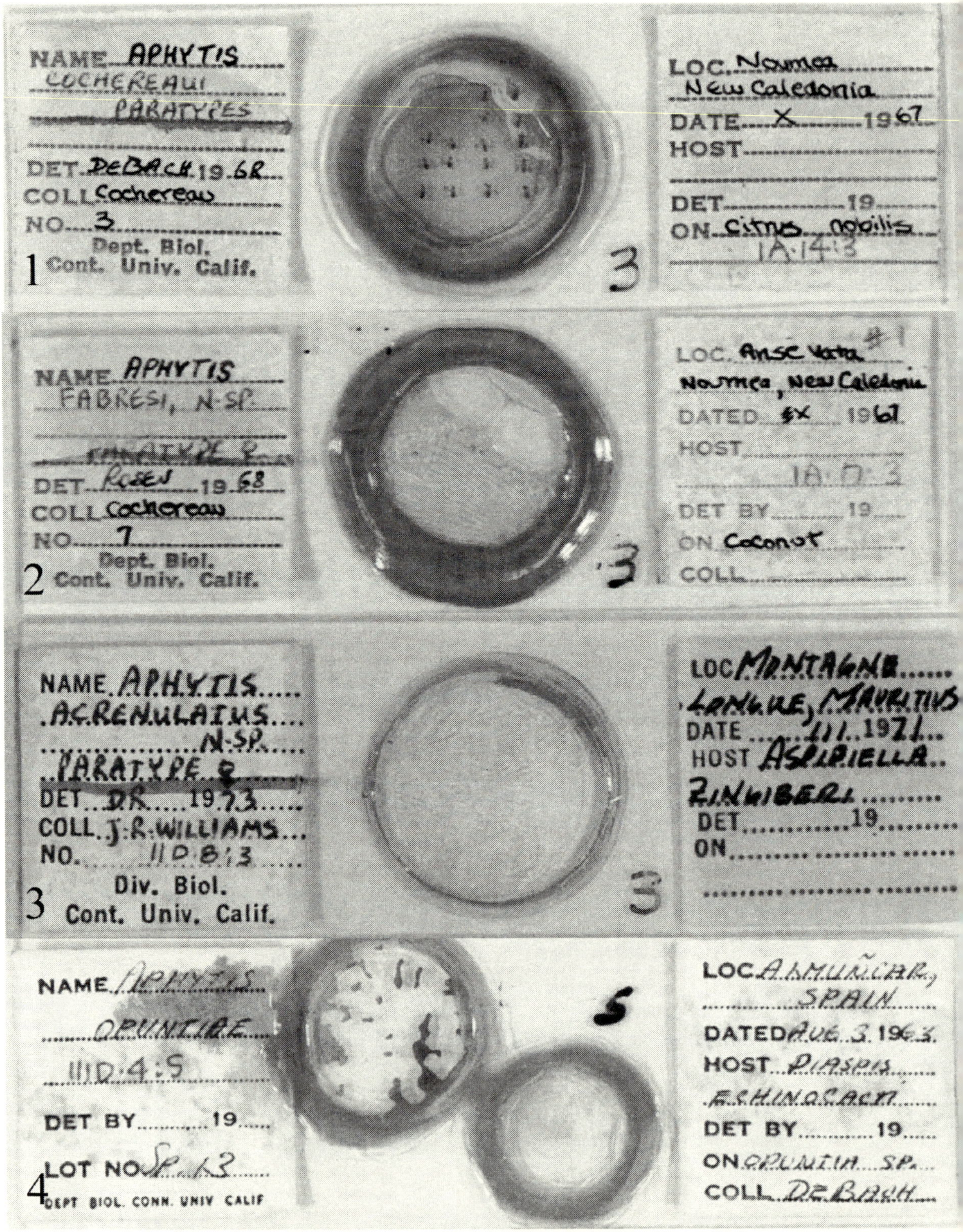

Figs 5-8. Major problems in UCR collection of slide-mounted *Aphytis* (prior to remounting). 5. Hoyer's deterioration on the slide with the original material of *A. melinus* (numerous specimens under three coverslips the coverslip ringed with Zut®, the mounting medium is darkened). 6. Hoyer's deterioration on the slide with the original material of *A. melinus* (numerous specimens under two coverslips ringed with Glyptal®, the mounting medium is darkened). 7. Hoyer's deterioration on the original syntype slide of *A. griseus* (numerous specimens under the same coverslip, the mounting medium is partially black and obscuring the specimens including the lectotype). 8. Hoyer's breakdown to a solid black color obscuring all the specimens on the original syntype slide of *A. africanus*.

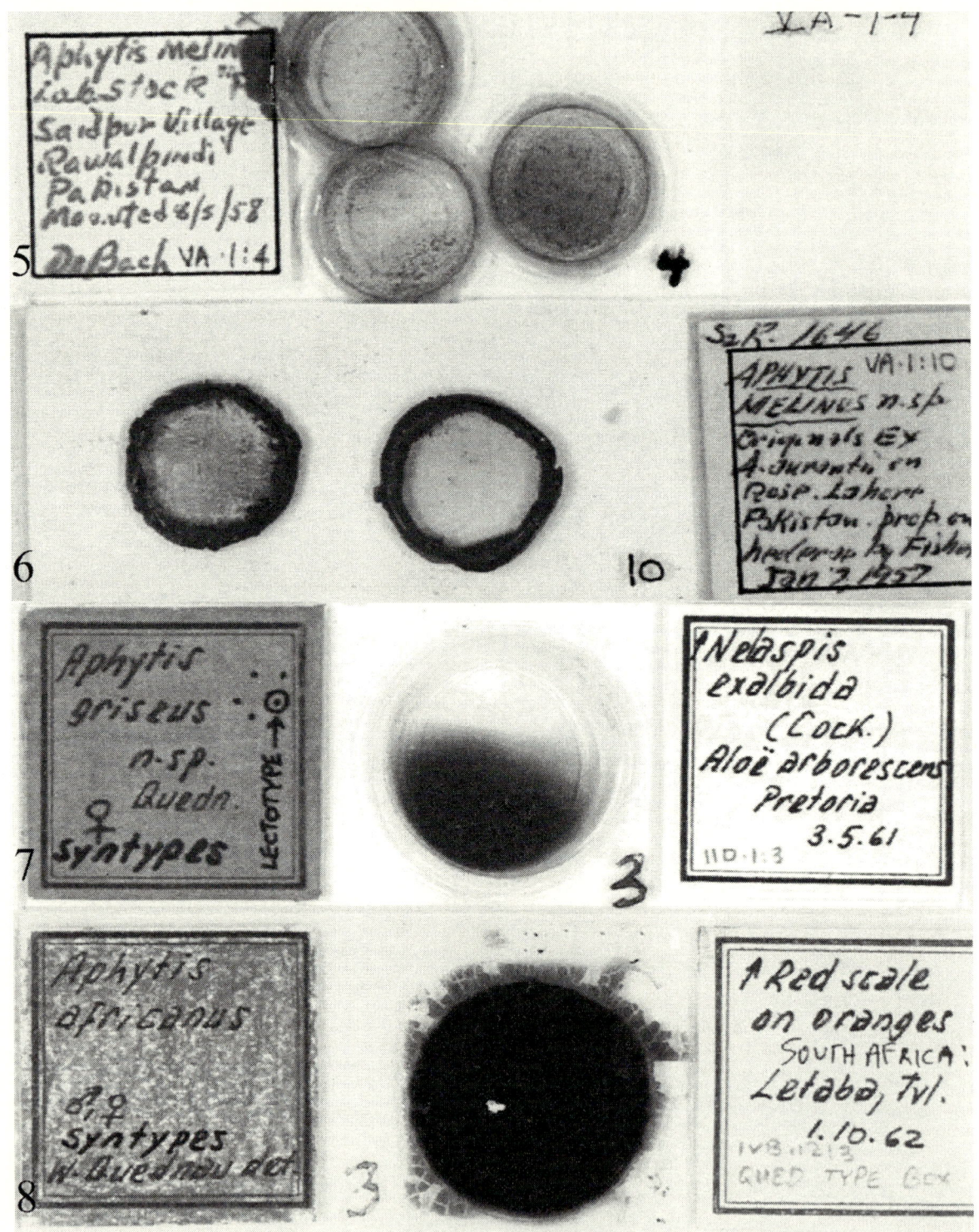

Figs 9-12. 9, 10. Major problems in UCR collection of slide-mounted *Aphytis* (prior to remounting). 9. Multiple specimens of *A. lingnanensis* mounted under the same coverslip ringed with Zut® (incomplete label information). 10. Multiple specimens and two different species mounted on a slide under the same coverslip (paratypes of *A. comperei* and non-type specimens of *A. hispanicus*). 11. The original syntype slide of *A. melinus* (from which the lectotype was selected) after remounting. 12. The lectotype slide of *A. melinus* after remounting.

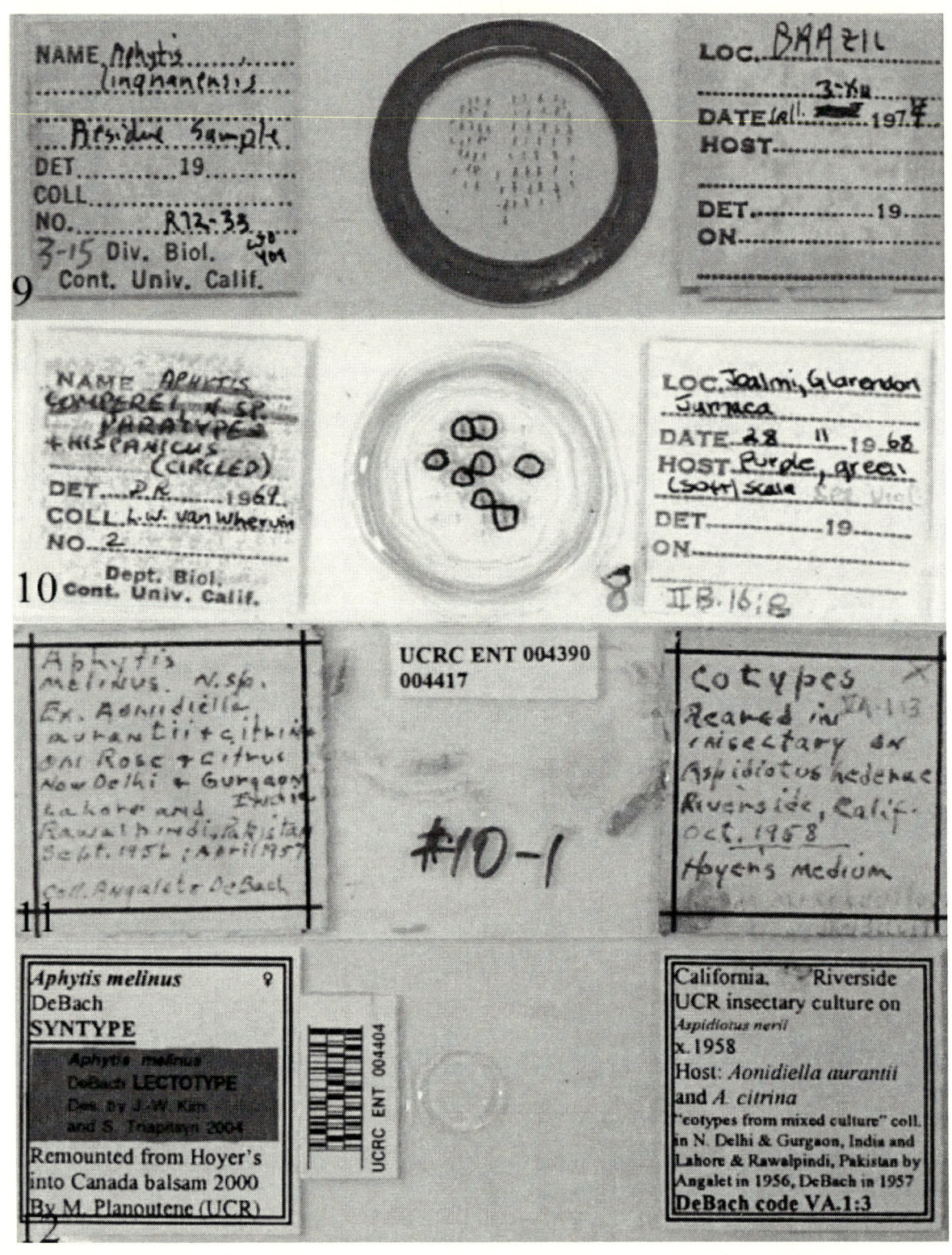